Studies in Applied Philosophy, Epistemology and Rational Ethics

Volume 43

Studies in Applied Philosophy, Epistemology and Rational Ethics (SAPERE) publishes new developments and advances in all the fields of philosophy, epistemology, and ethics, bringing them together with a cluster of scientific disciplines and technological outcomes: from computer science to life sciences, from economics, law, and education to engineering, logic, and mathematics, from medicine to physics, human sciences, and politics. It aims at covering all the challenging philosophical and ethical themes of contemporary society, making them appropriately applicable to contemporary theoretical, methodological, and practical problems, impasses, controversies, and conflicts. The series includes monographs, lecture notes, selected contributions from specialized conferences and workshops as well as selected Ph.D. theses.

Advisory Board

More information about this series at http://www.springer.com/series/10087

Woosuk Park

Philosophy's Loss of Logic to Mathematics

An Inadequately Understood Take-Over

 Springer

Woosuk Park
Humanities and Social Sciences
KAIST
Daejeon, Korea (Republic of)

ISSN 2192-6255   ISSN 2192-6263   (electronic)
Studies in Applied Philosophy, Epistemology and Rational Ethics
ISBN 978-3-030-06984-1   ISBN 978-3-319-95147-8   (eBook)
https://doi.org/10.1007/978-3-319-95147-8

This Springer imprint is published by the registered company Springer International Publishing AG
part of Springer Nature
The registered company address is: Gewerbestrasse 11, 6330 Cham, Switzerland

Preface

Essays originally published in Korean have been translated into English by me. I have made small changes, mainly stylistic, to most of the articles published internationally. In updating the book's chapter, I have striven for an overall consistency and coherence, I hope with satisfactory success. Supplementary readings are proposed in the Epilogue.

Chapter 2 first appeared in Korean in *Yonsei Philosophy*, 4, (1992), 53–80. Chapter 3 is a reprint of a paper published in *Modern Schoolman*, Vol. 67, (1990), 259–273. Chapter 4 appeared in *Logica Yearbook 2000*, (Prague: Czech Academy of Science), (2001), pp. 79–89. Chapter 5 appeared in *Korean Journal of Logic*, 11. 2, (2008), 1–57. Chapter 6 was published in *Erkenntnis*, 76 (3), (2012), 427–442. Chapter 7 was published in *Korean Journal of Logic*, 14.2, (2011), 1–37. Chapter 8 appeared in *Korean Journal of Logic*, 17 (1), (2014), 33–69. Chapter 9 was published as a discussion note in *Modern Schoolman*, Vol. 80, (2003), 144–153. Chapter 10 was published in *Korean Journal of Logic*, 9.2, (2006), 117–175. Chapter 11 was published in *Foundations of Science*, 21, (2016), 511–526. Chapter 12 was published (in English) in *Korean Journal of Logic*, 12.2, (2009), 141–170.

Daejeon, Korea (Republic of)   Woosuk Park

Acknowledgements

While I am indebted to all my teachers in logic and philosophy, I am especially grateful to Jorge J. E. Garcia and John Corcoran for their unfailing support. Nino Cocchiarella's detailed and much appreciated comments led to invaluable improvements of several of the book's chapters. As always, I have benefited from the criticism and encouragement of John Woods and Lorenzo Magnani.

Contents

1   Introduction   1
    References   6

Part I   The Fregean Legacy

2   Frege's Distinction Between "Falling Under"
    and "Subordination"   9
    1   Introduction   9
    2   The Distinction   10
    3   The Non-existence of the Distinction   12
        3.1   Traditional Logic   12
        3.2   19th Century Mathematicians   18
    4   Why Does the Distinction Matter to Frege?   23
        4.1   In Search of Logical Objects   24
        4.2   Frege's Analysis of the Statements of Number   25
        4.3   Frege's Function-Correlates   26
        4.4   The Distinction Between Falling Under and Subordination
              Versus the Distinction Between Class Membership
              and Class Inclusion   27
    References   28

3   Scotus, Frege, and Bergmann   31
    1   Cocchiarella's Thesis   33
    2   Scotus and Frege   34
    3   Bergmann and Frege   39
    4   Scotus and Bergmann   42
    References   44

4   On Cocchiarella's Retroactive Theory of Reference   47
    1   Introduction   47
    2   Predicable Concepts and Nominalized Predicates.   49
    3   Referential Concepts   51
         3.1   General Reference   51
         3.2   Singular Reference of Proper Names   52
    4   Frege and Cocchiarella   53
    5   Conclusion   55
    References   56

Part II   The Hilbert School

5   Zermelo and the Axiomatic Method   59
    1   Introduction   59
    2   The Discrepancy Between the Later Zermelo's Thought
         and Hilbert Program   62
    3   The Development of Hilbert's Thought on Axiomatic Method
         Between 1900 and 1917   63
         3.1   Implicit Definition in Hilbert   64
         3.2   Euclidean Axiom System and Hilbertian Axiomatic
                Method: Unending Frege/Hilbert Controversy   64
         3.3   The Application of the Axiomatic Method and Deepening
                the Foundations   67
         3.4   The Influence of Zermelo-Russell Paradox   71
    4   Zermelo's Position on the Axiomatic Method Around 1908   74
         4.1   The Motivations for the Axiomatization   74
         4.2   Zermelo's Reductionism   76
         4.3   The Problem of Selecting Axioms and the Axiom
                of Choice   78
         4.4   Could Zermelo's Axiomatic System of Set Theory
                Be the Implicit Definition of the Concept of Set?   82
    5   Concluding Remarks   84
    References   84

**6   Friedman on Implicit Definition: In Search of the Hilbertian
      Heritage in Philosophy of Science**   87
    1   Introduction   87
    2   Implicit Definition and the Relativized a Priori   89
    3   Carnap's Debt to Hilbert   94
         3.1   Hilbert on Deepening the Foundations   95
         3.2   Friedman on Carnap's Later Views on Theoretical Terms
                in Science   98
    4   Concluding Remarks   99
    References   100

7 Between Bernays and Carnap 103

1 Introduction 103

2 Hilbert's Axiomatic Method and Deepening the Foundations 106

3 The Limitation of Hilbert's Axiomatic Method from Bernays'
Point of View 108

    3.1 The Development of Axiomatic Method 108

    3.2 The Problem of Uniformity of the Axiomatic Method...... 112

4 Bernays' Criticism of Carnap 115

5 Concluding Remarks 117

References 118

Part III Goedel and Tarski

8 Patterson on Tarski's Definition of Logical Consequence 123

1 Introduction 123

2 Patterson's Interpretation of Tarski 128

    2.1 Intuitionistic Formalism and the Young Tarski 129

    2.2 Skepticism About Semantics 133

    2.3 Carnap and Tarski 136

    2.4 Carnap's and Tarski's Concepts of Logical Consequence ... 137

3 Beyond Patterson 140

References 143

9 On the Motivations of Gödel's Ontological Proof 145

1 Introduction 145

2 Back to Gödel's Original Proof 146

3 Gathering Clues 147

4 The Role of the Proof of the Existence of God in Mathematics
and Science... 150

5 Conclusion 153

References 153

Part IV Back to Aristotle

10 Ontological Regress of Maddy's Mathematical Naturalism 157

1 Introduction 157

2 Maddy as Realist.................................... 160

    2.1 Maddy's Strategy as a Realist to Meet Benacerraf's
Challenge.................................... 160

    2.2 The Analogy Between Science and Mathematics 161

    2.3 Aristotelian Aspects of Maddy's Realism.............. 166

    3   Maddy's Mathematical Naturalism: Beyond Quine and Gödel   166
        3.1   What Is Gone, What Is Newly Introduced,
              and What Is Revised.   166
        3.2   The Analogy Between Science and Mathematics
              Revisited   169
    4   Concluding Remarks: What Maddy's Belief Change Means   185
    References   185

11 **What if Haecceity Is not a Property?**   189
    1   Introduction   189
    2   The Common Assumptions About Haecceity   191
    3   Haecceity and the Identity of Indiscernibles   192
    4   *Haecceitas* from Scotus' Point of View   196
    5   What if Haecceity Is not a Property?   199
    6   Concluding Remarks   202
    References   205

12 **Biancani on *Scientiae Mediae***   207
    1   Introduction   207
    2   Wallace and Dear on the Motivations of Jesuit Debates
        on *Scientiae Mediae*   208
    3   The Chapter on *Scientiae Mediae*   210
    4   Biancani's Target   211
    5   Biancani on Perfect Demonstrations   212
    6   Biancani's Strategy   217
    7   Concluding Remarks   219
    References   220

Epilogue.   223

References   229

Chapter 1
Introduction

This is a collection of my papers on the history and philosophy of logic and mathematics published for the last thirty years. Virtually all the chapters tackle some particular logical, methodological, epistemological, and ontological issues that are not entirely clear in official history of modern logic. In retrospect, there were some very good reasons for me to be fascinated by the particular issues and the philosophers at the earlier stages of my research. Topically speaking, these chapters can be grouped under four parts. Part 1, which deals with Gottlob Frege, was motivated to understand what aspects of his logic were truly innovative in its revolution against the Aristotelian logic. Part 2 treats Hilbert and his associates and followers in the hope to understand the revolutionary change in the axiomatic method. Against that background, Part 3 discusses how to understand Tarski and Gödel as the towering figures whose problems are still with us. Finally, part 4 invokes some of the most influential positions in contemporary philosophy of mathematics. Both Maddy and Shapiro can be understood in terms of their reactions to Benaceraff's challenge to platonism in mathematics. Even though this part ends with a chapter on a renaissance philosopher, the main question is raised from our current situation in science and mathematics. Largely due to the foundational approaches in the first half of the twentieth century philosophy of mathematics, we do not fully understand the problems of application of mathematics. By introducing Biancani and Aristotelian philosophy of mathematics to the forefront, I want to hint at the urgent need to reconsider the Aristotelian position in logic and mathematics, which disappeared almost completely from the scene without good reasons in the early twentieth century.

There are already many textbooks in philosophy of mathematics and philosophy of logic, not to mention logic. The uniqueness of my book lies in its attempt to understand the philosophical problems raised in logic and mathematics in historical context. Rather than plunging into the unending disputes between different parties, I try to focus on the main issues that motivated the giants who established modern logic and philosophy of mathematics. By avoiding unnecessary technicalities of

© Springer International Publishing AG, part of Springer Nature 2018   1
W. Park, *Philosophy's Loss of Logic to Mathematics*, Studies in Applied Philosophy,
Epistemology and Rational Ethics 43, https://doi.org/10.1007/978-3-319-95147-8_1

logic and mathematics, it will be accessible to any reader who wants to understand modern philosophy of logic and mathematics.

Part 1, which deals with Gottlob Frege, was motivated to understand what aspects of his logic were truly innovative in its revolution against the Aristotelian logic. There are several surprising facts around this theme. First of all, Frege was a professional mathematician throughout his career. How was it possible for a mathematician to become the father of modern logic, analytic philosophy, and cognitive science? Secondly, we have reason to be surprised to witness the all too late revolution in logic. How was it possible for one man to revolutionize a field of scientific research dominated by the Aristotelian paradigm for more than two millenia? Finally, if Frege's reform is truly revolutionary, why has he not been well known to the intellectuals, if not to the general public? In other words, why the influence of modern logic has not been felt widely and unmistakably?

Frege frequently complains that others are ignorant of the distinction between "falling under" and "subordination". This criticism is not only directed against the philosophers who are under the influence of Aristotelian logic but also against the mathematicians of his time. In Chap. 2, I shall show that this distinction must be the vantage point for understanding Frege in both historical and philosophical contexts. Strangely, this distinction is not studied extensively nowadays. There are some good reasons for this. First, ironically, it is so well established as to become a triviality. Secondly, some people think that Frege's criticism of the aggregate view of sets is outdated. Consequently, we cannot understand why this distinction was so important to Frege. In what problem situation did Frege formulate this distinction? Were there any rival theories of predication? Was this distinction an ad hoc device for Frege in order to establish other important theses? What would happen if we lack this distinction? This chapter aims at a partial answer to these questions.

In Chap. 3, I shall compare the ontologies of Scotus, Frege, and Bergmann, thereby indicating the inseparable connection between logic and ontology. If N. Cocchiarella's recent discussion of Frege's function-correlate is correct, we have reason to assimilate Frege's ontology to the Avicennian-Scotistic tripartite ontology of individuals, universals, and common natures in themselves. Further, to the extent that Scotus' ontology is similar to Frege's ontology, we may have indirect evidence concerning how Bergmann would think about such an interpretation of Scotus' *haecceitas* ontology. I want to show that current treatments of individuation can be seriously challenged by the possible return of the common nature. My strategy will be as follows. In section A, I shall discuss Cocchiarella's thesis regarding Frege's function-correlates. In section B, by using Cocchiarella's thesis, I shall try to compare Frege's ontology with the Avicennian-Scotistic tripartite ontology. Some of the similarities and differences between these two ontologies will become clearer in the process. In section C, I shall examine Bergmann's interpretation of Frege's ontology. After having drawn attention to how Bergmann criticizes Frege's introduction of concept-correlates and value-ranges, we may understand, in section D, how Bergmann would view Frege's and Scotus' tripartite ontologies. Hopefully the peculiarity of Bergmann's theory of universals and his bare particular theory of individuation will stand out clearly against the background of Frege's and Scotus' ontologies.

Chapter 4 continues to get further insights from Cocchiarella's history and philosophy of logic in understanding the contrast of Aristotelian and Fregean logic. Recently Cocchiarella proposed a conceptual theory of the referential and predicable concepts used in basic speech and mental acts (Cocchiarella 1998). This theory is interesting in itself in that singular and general, complex and simple, and pronominal and nonpronominal, referential concepts are claimed to be given a uniform account. Further, as a fundamental goal of this theory is to generate logical forms that represent the cognitive structure of our speech and mental acts, as well as logical forms that represent only the truth conditions of those acts, it is an indispensable part of Cocchiarella's conceptual realism as a formal ontology for general framework of knowledge representation. In view of the recent surge of interest in his formal ontology by cognitive scientists and AI people, at least, Cocchiarella's theory of reference deserves careful examination Above all, however, the utmost value of Cocchiarella's theory of reference must be found in its challenge against what he calls "the paradigm of reducing general reference to singular reference of logically proper names" that pervades the 20th century (Cocchiarella 1998, 170). The aim of the present chapter is to provide an impressionistic sketch of Cocchiarella's challenge.

Part 2 treats Hilbert and his associates and followers in the hope to understand the revolutionary change in the axiomatic method. Even though it is almost common-sensical to cite the big three, i.e., logicism, formalism, and intuitionism, in the foundations of mathematics, through Russell and Wittgenstein, philosophers have paid much more attention to logicism compared to other schools in the foundations of mathematics. This is unfortunate, for one might give up the fine opportunity to secure a broader perspective on history of mathematics and the real practice of working mathematicians. There must be lots of lessons we can get from the detailed history of the interaction between mathematics and philosophy.

In this vein, Chap. 5 intends to examine the widespread assumption, which has been uncritically accepted, that Zermelo simply adopted Hilbert's axiomatic method in his axiomatization of set theory. What is essential in that shared axiomatic method? And, exactly when was it established? By philosophical reflection on these questions, we are to uncover how Zermelo's thought and Hilbert's thought on the axiomatic method were developed interacting each other. As a consequence, we will note the possibility that Zermelo, in his early as well as late thought, had views about the axiomatic method entirely different from that of Hilbert. Such a result must have far-reaching implications to the history of set theory and the axiomatic method, thereby to the philosophy of mathematics in general.

Encouraged by the recent surge of interest in logical positivism, I shall discuss in Chap. 6 the problem of the relationship between Hilbert and the logical positivism. To what extent and in what respects were logical positivists indebted to Hilbert? In particular, what exactly did they learn from the notion of implicit definition as the core of Hilbert's axiomatic method? In order to answer these questions, I shall first try to fathom Michael Friedman's mind about them. Secondly, I shall attempt to determine whether Carnap's approach to theoretical terms in science is nothing but an application of Hilbert's axiomatic method, as Bernays claims.

Since Bernays has not drawn scholarly attention that he deserves, in Chap. 7, it would be worth a while to continue to probe the questions raised in Chaps. 5 and 6. Only quite recently, the reevaluation of his philosophy, including the projects of editing, translating, and reissuing his writings, has just started. As a part of this renaissance of Bernays studies, this chapter tries to distinguish carefully between Hilbert's and Bernays' views regarding the axiomatic method. We shall highlight the fact that Hilbert was so proud of his own axiomatic method on textual evidence. Bernays' estimation of the place of Hilbert's achievements in the history of the axiomatic method will be scrutinized. Encouraged by the fact that there are big differences between the early middle Bernays and the later Bernays in this matter, we shall contrast them vividly. The most salient difference between Hilbert and Bernays will shown to be found in the problem of the uniformity of the axiomatic method. In the same vein, we will discuss the later Bernays' criticism of Carnap, for Carnap's project of philosophy of science in the late 1950s seems to be a continuation and an extension of Hilbert's faith in the uniformity of the axiomatic method.

Part 3 discusses how to understand Tarski and Gödel as the towering figures whose problems are still with us. They are not only the great logicians of all time but also the wonderful source of information to the history of logic in the twentieth century. I will hint at what kind of unconventional approaches might shed light on the history of logic in this regard.

Chapter 8 is basically an extensive review of Patterson's monumental book on Tarski's Philosophy of logic and language (Patterson 2012). We still do not know against what historical/philosophical background and motivation Tarski's definition of logical consequence was introduced, even if it has had such a strong influence. In view of the centrality of the notion of logical consequence in logic and philosophy of logic, it is rather shocking. There must be various intertwined reasons to blame for this uncomfortable situation. There has been remarkable progress achieved recently on the history of analytic philosophy and modern logic. In view of the recent developments of the controversies involved, however, we will have to wait years to resolve all these uneasiness. In this gloomy situation, Douglas Patterson's recent study of Tarski's philosophy of language and logic seems to have the potential to turn out to be a ground breaking achievement. This chapter aims at uncovering the state-of-the-art and fathoming the future directions of the research in this problem area by examining critically some unclear components of Patterson's study.

In recent years there has been a surge of interest in Gödel's ontological proof of the existence of God. In spite of all this extensive concern, it is not certain whether there is any improvement in understanding the motivations of Gödel's ontological proof. Why was Gödel so preoccupied with completing his own ontological proof? To the best of my knowledge, no one has dealt with this basic question seriously enough to answer it. In Chap. 9, I propose to examine Gödel's ideas against a somewhat larger background in order to understand his motivation for establishing the ontological proof. I shall point out that the value of Gödel's proof is to be found in the possible role of his proof of the existence of God in his philosophy as a whole as well as in its relative merit as an ontological proof. Hopefully, my guiding question as to Gödel's motivation will turn out to be extremely fruitful by enabling us to fathom his mind regarding God and mathematics. After all, we are interested in Gödel's ontological proof *because* it a proof presented by the great mathematician. At the same time, we may fathom how Gödel understood the axiomatic method indirectly from his ontological proof.

Part 4 invokes some of the most influential positions in contemporary philosophy of mathematics. Why am I counting Penelope Maddy and Stewart Shapiro as representing the most influential positions? Well, in view of the fact that it is simply impossible to do justice to all philosophies of mathematics in contemporary philosophy, selection is unavoidable. So, my choice must have been made by what I believe to be the most promising direction for the future philosophy of mathematics, i.e., Aristotelian philosophy of mathematics. Maddy and Shapiro have more than enough aspects to be discussed under the general rubric of Aristotelian mathematics.

Chapter 10 is an attempt to probe the question as to why Maddy gave up mathematical realism and moved to her own version of mathematical naturalism. According to one widespread hypothesis, Maddy's change of mind was brought up by her criticism of Quine-Putnam indispensability argument. Though quite convincing, it is not good enough to explain why one has to give up mathematical realism. The analogy of science and mathematics will instead be shown to be the better perspective to fathom Maddy's changing beliefs. For this purpose, we have to understand to what extent Maddy's thought in her realist years, which was strongly influenced by Quine and Gödel, was governed by the analogy between science and mathematics. Also, we have to understand why Maddy gave up the analogy, and thereby gave up mathematical realism. Finally, some criticisms against Maddy's abandonment of the analogy will be examined so as to hint at the reasons why I believe Maddy's intellectual journey in mathematical ontology is rather regress than progress.

In some sense, both ontological and epistemological problems related to individuation have been the focal issues in the philosophy of mathematics ever since Frege. However, such an interest becomes manifest in the rise of structuralism as one of the most promising positions in recent philosophy of mathematics. The most recent controversy between Keränen and Shapiro seems to be the culmination of this phenomenon. Rather than taking sides, in Chap. 11, I propose to critically examine some common assumptions shared by both parties. In particular, I shall focus on their assumptions on (1) haecceity as an individual essence, (2) haecceity as a property, (3) the classification of properties, and thereby (4) the search for the principle of individuation in terms of properties. I shall argue that all these assumptions are mistaken and ungrounded from Scotus' point of view. Further, I will fathom what consequences would follow, if we reject each of these assumptions.

We can witness the recent surge of interest in the controversy over the scientific status of mathematics among Jesuit Aristotelians around 1600. Following the lead of Wallace, Dear, and Mancosu, I propose to look into this controversy in more detail. For this purpose, I shall focus on Biancani's discussion of *scientiae mediae* in his dissertation on the nature of mathematics. From Dear's and Wallace's discussions, we can gather a relatively nice overview of the debate between those who championed the scientific status of mathematics and those who denied it. But it is one thing to fathom the general motivation of the disputation, quite another to appreciate the subtleties of dialectical strategies and tactics involved in it. It is exactly at this stage when we have to face some difficulties in understanding the point of Biancani's views on *scientiae mediae*. Though silent on the problem of *scientiae mediae*, Mancosu's discussions of the Jesuit Aristotelians' views on *potissima* demonstrations, mathematical explanations, and the problem of cause are of utmost importance in this regard, both historically and philosophically. In Chap. 12, I will carefully examine and criticize some of Mancosu's interpretations of Piccolomini's and Biancani's views in order to approach more closely what was really at stake in the controversy.

Largely due to the foundational approaches in the first half of the twentieth century philosophy of mathematics, we do not fully understand the problems of application of mathematics. By introducing Biancani and Aristotelian philosophy of mathematics to forefront, I want to hint at the urgent need to reconsider the Aristotelian position in logic and mathematics, which disappeared almost completely from the scene without good reasons in the early twentieth century. In the Epilogue, I will hint at how to do more interesting and exciting researches for such a project.

References

Cocchiarella, N. (1998). Reference in conceptual realism. *Synthese, 114,* 169–202.
Patterson, D. (2012). *Alfred Tarski: Philosophy of language and logic.* U.K.: Palgrave Macmillan.

Part I The Fregean Legacy

Chapter 2
Frege's Distinction Between "Falling Under" and "Subordination"

Abstract Frege frequently complains that others are ignorant of the distinction between "falling under" and "subordination". This criticism is not only directed against the philosophers who are under the influence of Aristotelian logic but also against the mathematicians of his time. I shall show that this distinction must be the vantage point for understanding Frege in both historical and philosophical contexts. Strangely, this distinction is not studied extensively nowadays. There are some good reasons for this. First, ironically, it is so well established as to become a triviality. Secondly, some people think that Frege's criticism of the aggregate view of sets is outdated. Consequently, we cannot understand why this distinction was so important to Frege. In what problem situation did Frege formulate this distinction? Were there any rival theories of predication? Was this distinction an ad hoc device for Frege in order to establish other important theses? What would happen if we lack this distinction? This chapter aims at a partial answer to these questions.

Keywords Aristotle · "Falling under" · Frege · Predication · Subordination

1 Introduction

Frege frequently complains that others are ignorant of the distinction between "falling under" and "subordination" (Angelelli 1967, p. 107, 125n). This criticism is not only directed against the philosophers who are under the influence of Aristotelian logic but also against the mathematicians of his time (Jourdain 1912, 204; Angelelli 1967, 107–108; Frege 1893 in Geach and Black 1952, 149). Isn't

---

An earlier much shorter version was published in Korean as Park (1992), and also reprinted as Chap. 3 of Park (1997). But this published version completely deleted Sect. 3.1 of the original version, which was submitted as a final term paper to John Corcoran's graduate seminar entitled "History of Logic, Spring 1986, State University of New York at Buffalo". Such a cut was rather drastic, for deleted part had dealt with the lack of the distinction between "falling under" and "subordination" in traditional logic up until the 19th century. In this paper, I shall recover it entirely. Also, I shall attempt a thorough update by incorporating more recent advances in Frege scholarship.

© Springer International Publishing AG, part of Springer Nature 2018
W. Park, *Philosophy's Loss of Logic to Mathematics*, Studies in Applied Philosophy, Epistemology and Rational Ethics 43, https://doi.org/10.1007/978-3-319-95147-8_2

this the hallmark of Frege? As long as we honor Frege as the founder of modern symbolic logic, and as long as we mention the reunion of logic and mathematics in order to prove the superiority of new logic, this distinction must be the vantage point for understanding Frege in both contexts, i.e., in the history of philosophy and the foundations of logic and mathematics.

Strangely, this distinction is not studied extensively nowadays.[1] There are some good reasons for this. First, ironically, it is so well established as to become a triviality. This distinction is rather given to many, and they tend to think that there must be predecessors who anticipated it. Secondly, some people think that Frege's criticism of the aggregate view of sets is outdated (Resnik 1980, p. 206). According to them, we have a new aggregate view of sets which is immune to Frege's criticism, i.e., the iterative concept of set (Boolos 1971, 486–487). In other words, Frege's criticism of 19th century mathematicians' failure to draw the distinction is no longer interesting.

Consequently, we cannot understand why this distinction was so important to Frege. In what problem situation did Frege formulate this distinction? Were there any rival theories of predication? Was this distinction an ad hoc device for Frege in order to establish other important theses? What would happen if we lack this distinction? This chapter aims at a partial answer to these questions. In Sect. 2, the distinction between "falling under" and "subordination" is briefly introduced. In Sect. 3.1, I shall examine the hypothesis that in Aristotelian logic predication was interpreted as Fregean "subordination" together with some possible objections to it. In Sect. 3.2, I shall discuss how Frege criticized his contemporary mathematicians for their failure to draw the distinction. In Sect. 4, I shall fathom why the distinction was so important to Frege.

2 The Distinction

Perhaps the most elaborate presentation of the distinction can be found in Frege (1892) (Geach and Black 1952, p. 51).[2] Here he began from distinguishing an object's falling *under* a first-level concept from a concept's falling *within* a

---

[1]Ignacio Angelelli and Nino B. Cocchiarella seem to be rare exceptional cases. Angelelli devotes a chapter on this distinction, and scrutinizes the question why it was lacking in traditional philosophy. As a consequence, this chapter is full of interesting information. However, in general, he tends to presuppose the correctness of the distinction rather than contesting its merit critically. He never presents clearly what the distinction itself is. Nor does he examine Frege's criticism of 19th century mathematicians in detail. For example, he never quotes from Dedekind's own works (Angelelli 1967). Cocchiarella's analysis is full of logical and philosophical insights. Since he deals with all important issues related to Frege, however, it may not be a kind introduction to the distinction between "falling under" and "subordination" for beginners or general public.

[2]The distinction is of early origin in the development of Frege's thought. In several places we can find it out. For example, at the beginning of §53, he drew the distinction between properties of a concept and the mark of the concept. And, immediately he added the following explanation:

second-level concept. From this he concludes that the distinction of concept and object holds with all its sharpness. Then follows the distinction between 'property' and 'mark'. In order to contrast them, he resorted to the distinction between "falling under" and "subordination". I quote an entire paragraph, for this is our key text:

> I call the concepts under which an object falls its properties; thus
>> 'to be Φ is a property of Γ'
> Is just another way of saying;
>> 'Γ falls under the concept of aΦ'.
> If the object Γ has the properties Φ, X, and Ψ, I may combine them into Ω; so that it is the same thing if I say that Γ has the property Ω, or, that Γ has the properties Φ, X, and Ψ. I then call Φ, X, and Ψ marks of the concept Ω, and at the same time, properties of Γ. It is clear that the relations of Φ to Γ and to Ω are quite different, and that consequently different terms are required. Γ *falls under* the concept Φ; but Ω, which is itself a concept, cannot fall under the first-level concept Φ; only to a second-level concept could it stand in a similar relation. Ω is, on the other hand, *subordinate* to Φ. (Frege 1892, p. 51. Emphases are mine)

In the same place, he used the following example. We may say that 2 is a positive whole number less than 10.

> Here
>> to be a positive number (Φ),
>> to be a whole number (X),
>> to be less than 10 (Ψ),
> appear as properties of the object 2, and also as marks of the concept
>> positive whole number less than 10 (Ω).
> This is neither positive, nor a whole number, nor less than 10. It is indeed subordinate to the concept *whole number*, but does not fall under it.

By using standard notation of modern symbolic logic, we can understand it as follows:

> 2 falls under being a whole number. X_2
> Being a positive whole number less than 10 is subordinate to being a whole number. $\forall x$ ($\Omega x \rightarrow Xx$)

In other words, the distinction between "falling under" and "subordination" is closely related to the distinction between singular propositions and universal

---

"These latter (the characteristics of which make up the concept) are properties of things which fall under the concept, not of the concept" (Frege 1884, 64$^e{}_7$). Moreover, at the end of the section, he made it clear that we should not confuse the relation of one concept's falling under (within) another higher concept with the subordination of species to genus (Frege 1884, 65$^e{}_7$).

propositions.[3] What a trivial distinction! Surprisingly, however, this trivial distinction had been lacking until Frege established it. From the history of logic, we hear that traditional logicians (in particular, in the post-Renaissance period) interpreted singular propositions as A-type propositions, i.e., universal affirmative propositions (Copi 1982, p. 239; Prior 1963, 160).[4] Also, we can see how the pioneers of modern symbolic logic were so proud of this innovation.

Russell reported that the enlightenment he derived from Peano had come from two purely technical advances. He did not forget to indicate the fact that both of these advances had been made by Frege. And, these two alleged advances were the two sides of the very distinction between "falling under" and "subordination":

> The first advance consisted in separating propositions of the form 'Socrates is mortal' from propositions of the form 'All Greeks are mortal'.... The second important advance that I learnt from Peano was that a class consisting of one member is not identical with that one member. (Russell 1959, p. 52)

Then, first, we should confirm whether there had been no such distinction in philosophy as well as in mathematics.

3 The Non-existence of the Distinction

3.1 Traditional Logic

3.1.1 Ambiguity of 'be'

It is a commonsense that there is ambiguity of the verb 'be' in Indo-European languages. We usually distinguish 'is of predication' and 'is of identity'. Also, we might indicate the usage of 'is of existence'. Of course, Frege understood the distinction clearly. He distinguished between "'is' as a mere copula, as in the proposition 'The sky is blue'" and "'is' which has the sense of 'is identical with' or 'is the same as' (Frege 1884, 69e). Some people even emphasize Frege's distinction between the meaning of the verb 'be' as "one of the most widely accepted inventions of his and an

---

[3]As John Corcoran indicates, the idea that singulars are logically equivalent to universals is not the same as the idea that "falling under" is the same as "subordination". Following Quine, we can treat Ps as logically equivalent to $\forall x(x = s \supset Px)$. Even among the schoolmen we can find a thinker who interpreted "Socrates is running" as "Everything that is Socrates is running" (Prior 1963, p. 160). In this respect, it is by no means clear why traditional logicians (in particular in the post-Renaissance period) treated singular propositions as universal propositions. If it is the result of their theory which considers "falling under" as "subordination" (i.e., as will be examined in the next section), then it is still more dubitable why it was treated that way more clearly in post-Renaissance period.

[4]As Cocchiarella points out, here I am leaving unexplained how traditional logicians took predication to function in E, I, O type propositions (Cocchiarella 2015b).

integral part of most current treatments of semantics" (Hintikka 1979, p. 72).[5] However, such an insight cannot be said a breakthrough. Some people say that even Plato clearly distinguished the various meanings of the verb 'be' (Ackrill 1957).[6]

Then, there seems to be no hope for tracing the reason why traditional logicians failed to draw the distinction between "falling under" and "subordination" from the distinction between "is of predication" and "is of identity". Were they simply ignorant of the subdivision of predication into "falling under" and "subordination"? There is no doubt that the ambiguity of the verb 'be' plays a significant role in making matters complicated in predication theory, old and new. But it is not easy to pin down exactly what is wrong.

3.1.2   Aristotle: "Falling Under" as "Subordination"

As is well-known, in his *Categories* 5.2, Aristotle treated genera and species as secondary substances. According to Kneale and Kneale, this blurs "the all-important distinction between singular and general propositions" (Kneale and Kneale 1962, p. 31). In Aristotelian framework, the proposition "Man is an animal" is the paradigm case of predication. In Fregean language, this amounts to saying "the class of man is subordinate to the class of animals". Thus far, there is no problem. But, while Frege would interpret "Socrates is a man" and "Socrates is an animal" as exemplifying "falling under", Aristotle would treat them as cases of subordination: "This individual man is a man", and "This individual man is an animal", respectively. In the former, an individual man is subordinate to a lowest species. In the latter, only derivatively can we say "An individual man is an animal", because the class of man is subordinate to the class of animals:

> Whenever one thing is predicated of another as of a subject, all things said of the subject also. For example, man is predicated of the individual man, and animal of man; so animal

---

[5]The following quote from Hintikka (1981) seems extremely informative: "The sharpest specific difference between Frege's logical notation and the ideas of his predecessors lies in his treatment of verbs for being. Such verbs are, according to Frege and his followers, ambiguous in that they have to be translated into the logical notation in at least four different ways: (i) by the identity sign '=' (the 'is' of identity); (ii) by the existential quantifier (the 'is' of existence); (iii) by predicative juxtaposition (the 'is' of predication or the copula), and (iv) by a general implication (the 'is' of class inclusion). In 1914 Bertrand Russell (1914) called this fourfold distinction 'the first serious advance in real logic since the time of the Greeks', and it has been incorporated in all the usual systems of first-order logic (lower predicate calculus, quantification theory). Hence everybody who has been using the notation of first-order logic for the purpose of semantical representation is committed to the ambiguity of 'is'. This applies to linguists and philosophers otherwise as unlike each other as George Lakoff, Noam Chomsky, W. V. Quine, Donald Davidson, and Ludwig Wittgenstein. Frege's distinction has thus become one of the most widely accepted inventions of his and an integral part of most current treatments of semantics (Hintikka 1981, p. 72).

[6]Corcoran informed me of the fact that Russell cites De Morgan (1847, 49–53) instead of Frege (Russell 1903, 64n).

will be predicated of the individual man also- - -for the individual man is both a man and an animal. (Aristotle, *Categories*, 3, 1b 10, Ackrill 1963)

From these facts we might formulate a working hypothesis that Aristotle treated "falling under" as "subordination", even though we cannot be sure whether Aristotle was simply ignorant of the distinction between "falling under" and "subordination", or schemingly assimilated "falling under" to "subordination". In the following, I shall test this hypothesis.

3.1.3   Predication *in Quid* and *in Quale*

Against the hypothesis that Aristotelian logic treats any predication as subordination in Fregean sense, a doubt naturally arises about whether the distinction between predication *in quid* and *in quale* is a counter-example. Can't we pair "falling under" with *in quale* on the one hand, and "subordination" with *in quid* on the other hand? Let us check whether this works.

We can find a crystal-clear formulation of the distinction between *in quid* and *in quale* in Porphyry (1975, p. 8, Porphyry 1966). Of course, we can go back to Aristotle.[7] Be that as it may, according to these old logicians, predication *in quid* is in answer to the question "What is it?", while predication *in quale* is in answer to the question "Of what sort is it?". So, they distinguished (1) "Socrates is a man" from (2) "Socrates is white", and distinguished (3) "Man is an animal" from (4) "Man is mortal".

At first sight, the example sentence of predication *in quale* seems to be also a good example of "falling under": (2) "Socrates is white". If it were the case that only singular propositions are treated as *in quale*, then our attempt to assimilate *in quale* to "falling under" would be successful. Unfortunately, however, the sentence (3) "Man is an animal" is also an example of *in quale*, and in this example a lower species is subordinate to a higher genus. Now the old logicians find out a lower thing's subordination to a higher species in singular propositions such as (2) "Socrates is white":

> Differentia is predicated of species and individuals .... Porphyry is predicated of the species of which it is the property and of the individuals under the species. And accident is predicated of species and individuals. (Porphyry 1975, p. 21, Porphyry 1966)

We may further demonstrate the point that, in *in quale* as well as in *in quid*, lower things (or species) is subordinate to a higher species (or genus) and thus that in both cases we can see Fregean "subordination" by paraphrasing the examples given above into old logicians' awkward language of addition as follows:

> (1') This snubnosed white *man* called Socrates is a *man*.

> (2') This snubnosed *white* man called Socrates is *white*.

---

[7]This distinction must originate from Aristotle. As a matter of fact, in *Topics* 102a 32, we can find Greek equivalent of "in eo quod quid est". Interestingly enough, I am unable to find out Aristotle's equivalent of "in eo quod quale est".

(3') The rational mortal risible *animal* called man is an *animal*.

(4') The rational *mortal* risible animal called man is *mortal*.

From all these, we can safely conclude that the function of the distinction between predication *in quid* and *in quale* is totally different from that of Frege's distinction between "falling under" and "subordination". Our hypothesis is still working.

3.1.4 Collective and Distributive Predication

One more powerful objection to our hypothesis comes from the fact that there is a distinction between collective predication and distributive predication in traditional logic.[8] Traditionally, discussion of informal fallacies has been a part of introductory logic texts. In such discussions, we always meet the fallacy of composition (and division). Copi gives us the following example as a fallacy of composition:

> A bus uses more gasoline than an automobile.
>
> Therefore all buses use more gasoline than all automobiles. (Copi 1982, p. 125)

He goes on to say that this version of the fallacy "turns on a confusion between the 'distributive' and the 'collective' use of general terms". Accordingly, "buses use more gasoline than automobiles, distributively, but collectively automobiles use more gasoline than buses, because there are so many more of them".

We can trace the first appearance of this fallacy back to Aristotle (Aristotle, *Soph. Ref.* 166 a23; See Hamblin 1970, 20). Moreover, according to Quine, the distinction between 'class-membership' and 'class-inclusion'—which parallels the distinction between "falling under" and "subordination"—existed in traditional logic in a *tentative form* as a distinction between "distributive" and "collective predication", drawn to resolve the fallacies of composition and division. His example is as follows:

> Peter is an Apostle.
>
> The Apostles are twelve.
>
> Therefore Peter is twelve. (Quine 1940, p. 189)

Now, we can summarize the discussion this way:

1. The fallacy of composition was already familiar to Aristotle.
2. The fallacy of composition is closely related to the distinction between distributive and collective predication.

---

[8]Corcoran kindly reminded me of the relevance of this distinction to my discussion. He was careful enough not to identify the distinction with the fallacy of composition. The late Charles Lambros also helped me in various ways. He was so critical of Frege that he even thought that Frege (1884) would not have argued if he had known the distinction between collective and distributive predication: "Frege's error here is glossing over the now familiar distinction between distributive and collective predicates" (Lambros 1976, 381).

3. The distinction between distributive and collective predication is the anticipation of the distinction between "falling under" and "subordination".
4. Therefore, the distinction between "falling under" and "subordination" was known to traditional logicians.

This argument is extremely plausible. And it must be an arduous task to counter this argument. Then, should we say that Frege's distinction was at best a clear formulation of what had been familiar to logicians? I think not. No matter how plausible this argument might be, we also have some ground to criticize this argument. The first premise may be fully conceded. For the other premises and the conclusion, however, there is room for qualification.

First of all, it is not clear what relationship holds between the fallacy of composition and the distinction between distributive and collective predication. Is this distinction a necessary condition of the fallacy of composition? Absolutely not! As Copi indicates, this distinction is related to only one of the two kinds of the fallacy of composition. This point can be strengthened even more. In a footnote to William of Sherwood's *Introduction to Logic*, Kretzmann wrote:

> The Aristotelian-medieval doctrine of the fallacies of composition and division is much broader than that found in textbooks of the modern period, which usually recognize only the difficulties associated with the compounded and divided senses of generalizations, such as 'all the interior angles of a triangle equal 180°'. (William of Sherwood 1966, p. 140, n. 49)

From this quotation, it becomes clear that the special kind of fallacy of composition related to distributive and collective distinction is by no means the representative of the fallacy of composition.

However, let us assume that traditional logicians were alert enough to discern this special kind of the fallacy of composition. Even now we cannot infer that they drew the distinction between distributive and collective predication. At least we need some historical evidence. Moreover, even if we concede that they drew the distinction, we cannot immediately conclude that they have a predication theory which is armed with the distinction between "falling under" and "subordination".

Let's go back to Quine's example. Should we read "The apostles are twelve" collectively or distributively? Perhaps, it is best read collectively, but, please note that this statement is a statement of number whose form Frege was anxious to reveal. Is this an example of "class-membership" or "class-inclusion"? Isn't Quine begging the question here? Please look at the following example:

(1) Every natural number is finite (distributive reading) - - - true
(2) The set of natural numbers is finite (collective reading) - - - false

It seems to me (1) is rather an example of "subordination", while (2) is that of "falling within". I don't know how Quine would have read these. At least, however, it is not clear how we can compare the distinction between collective and distributive predication to the distinction between "falling under" and "subordination", or to the distinction between "class-membership" and "class-inclusion".

If it is not easy to match the distinctions, we might go one step further to guess that traditional logicians could distinguish collective predication and distributive predication without knowing the difference between "falling under" and "subordination". In other words, they could distinguish them by interpreting all predications as "subordination". Perhaps, it would not be so difficult to paraphrase the example sentences in terms of "subordination".

Finally, we should note that even if they did make the distinction between collective and distributive predication on the basis of a distinction which is a counterpart of Frege's distinction between "falling under" and subordination", it has not been so influential in the history of logic. As I already indicated, after the post-medieval period logicians tended more to interpret singular propositions as universal propositions. If the distinction, if there was any, had been influential, such a phenomenon would not have occurred.

Therefore, even though we have to concede the close relationship between the distinction between collective and distributive predication with Frege's distinction, by now it would be irresponsible to say that the former anticipates the latter.

3.1.5 Subsumption as "Subordination"

Nowadays it is common to use the word "subsumption" as a synonym of "falling under". For example, Kluge states as follows:

> An object is said to *fall under* or *be subsumed under* a concept if and only if it has those properties that are characteristics of the concept.... This relationship of *subsumption* holds only between first-level concepts and objects. (Kluge 1971, p. x)

However, we can see that Kant used the term "subsumption" frequently. So, one might think that this is a counter-example to our hypothesis that pre-Fregeans (typically, Aristotelian logicians) interpreted predication as subordination exclusively.

However, Kant seems to use the term "subsumption" not in the sense of "falling under" but in the sense of "subordination". Kant wrote, for example:

> How, then, is the *subsumption* of intuitions under pure concepts, the *application* of a category to appearances, possible? (B177/A138, Kant 1871, 1879). (Kant's emphasis)

If we do not bear in mind that he meant "subordination" by the term "subsumption", his way of speaking seems to be committing solecism in the Kantian language of predication, for a category cannot be directly applied as predicates to objects. It is impossible to subsume an intuition under a pure concept. The following passage could be a clue to clear understanding of what he meant by "subsumption":

> In all subsumptions of an object under a concept the representation of the object must be *homogeneous* with the concept.... But pure concepts of understanding being quite heterogeneous from empirical intuitions.... can never be met with in any intuition. (A137/B176)

As a matter of fact, some third thing, i.e., schematism was needed for Kant in order to bridge the gulf between pure concepts and intuitions:

Obviously there must be some third thing, which is homogeneous on the one hand with the category, and on the other hand with the appearance, and which thus makes the application of the former to the latter possible. (B177/A138)

If Kant did not mean by "subsumption" Freagean "falling under", the only alternative is "subordination". At least, only when we interpret Kant's term "subsumption" as "subordination", we can read Kant consistently.

Interestingly, Schröder, in Chap. 1, Sect. 1 of his *Algebra der Logik* entitled "subsumption" cited the following sentences as the typical example of subsumption:

"a ist b"
"alles a ist b" (Schröder 1890, p. 133)

As was said, universal affirmative propositions are examples of "subordination" in Frege, so, it is clear that Schröder did not use the term "subsumption" exclusively in the sense of "falling under". The point is enough for my present purpose. So, thus far, our initial hypothesis that traditional logicians treated predication as "subordination" has survived these tests.

3.2   19th Century Mathematicians

3.2.1   Membership and Class Inclusion

In Frege (1893) (Greach and Black 1952, p. 149), Frege explicitly indicated Dedekind's neglect of the distinction between "falling under" and "subordination":

Indeed what Dedekind really means when he calls a system a 'part' of a system is that a concept is subordinated to a concept or an object falls under a concept.[9]

According to Frege, Dedekind and Schröder shared a common mistaken view because of which they fail to see the distinction.[10] As a matter of fact, Frege criticized Schröder for this failure in his review of Schröder:

---

[9]Of course, by 'is a part of' Dedekind meant 'is a subset of', i.e., class inclusion. So, one might think that Frege was being too fast here. It seems that Frege thought that the distinction between "class inclusion" and "class membership" is not different from the distinction between "subordination" and "falling under": "We must keep separate from one another: (a) the relation of an object (an individual) to the extension of a concept when it falls under the concept (the *subter* relation); (b) the relation between the extension of one concept and that of another when the first concept is subordinate to the second (the *sub* relation) (Geach and Black 1952, p. 106). I will deal with this point in more detail below.

[10]For a rather detailed examination of Frege's criticism of Schröder, see Cocchiarella (1987, pp. 89–90).

The important thing here is that our attention is drawn to the essential difference between two relations for which the author uses the same sign (the sign of 'eventuelle subordination' or inclusion)....

In order to bring clarity into the matter, it will be necessary to correct Schröder's mistake, and, when things are different, to use different signs for them. So we lay it down that

'A sub B'

is to assert that A is a class subordinate to the class B. On the other hand,

'A subter B'

is to express A's being comprised as an individual under the class B. (Frege 1895, in Geach and Black 1952, p. 92, 93)

Before critically examining whether we should draw such a distinction as Frege urges, it must be our duty to check whether that distinction is really lacking in Dedekind, Schröder, or any other 19th century mathematicians.

Dedekind, whether he was ignorant or deliberate, didn't distinguish "falling under" from "subordination". He used the symbol '∋' for indicating class inclusion and membership indifferently:

A system A is said to be *part* of a system S when every element of A is also element of S. Since this relation between a system A and a system S will occur continually in what follows, we shall express it briefly by the symbol A∋S....

Since further every element s of a system S by (2) can be itself regarded as a system, we can hereafter employ the notation s∋S.

As a consequence of this lack of the distinction, Dedekind fails to distinguish the property of a set per se from the property of the elements of the set. In other words, he didn't know the levels of concepts (Dedekind 1872).

In order to prove that Schröder is ignorant of the distinction between "falling under" and "subordination", Frege quoted some passages from Schröder's book:

And one individual may likewise be termed a class- - -one that contains only this individual itself....But also any class that itself contains many individuals can again be represented as an object of thought and accordingly as an individual (in a broader sense, e.g. as a 'relative' individual in regard to higher classes) (Frege 1895; Geach and Black 1952, p. 87)[11]

As in Dedekind, there is no distinction between class membership and class inclusion.

---

[11]"Und auch *ein* Individuum mögenwir bezichnen als eine Klasse, welche eben nur dieses Individuum selbst enthält. Ein jedes gedankending kann zu solchen Individuum gestempelt warden....Auch jene Klasse aber, die selber eine Menge von Individuen umfasst, kann wieder al sein Gedankending und demgemäss auch al sein "Individuum" (im weiteren Sinne, z.B. "relativ" in Bezug auf höhere Klassen) hinfestellt warden (Schröder 1890, 148).

3.2.2   The Alleged Consequence of the Lack of the Distinction

At the beginning of his critical review of Schröder's book, Frege made it clear why he intended to scrutinize Schröder's point of view. It is to make apparent the necessity of a distinction that seems to be unknown to many logicians (Geach and Black 1952, p. 86). The distinction he was referring to must be the distinction between "falling under" and "subordination", as is clear from his own summary of his discussion in the review (Geach and Black 1952, p. 106). Moreover, he treated this topic in the introduction of Frege (1893) (Geach and Black 1952, pp. 148–151). Thus it is evident that Frege himself endorsed this distinction and heavily depended on it. Then we should raise the following question: Why is this distinction so important? What kind of disaster might result, if we ignore it, as Dedekind and Schröder did?

Frege would answer these questions as follows. Because an empty class should not occur if we follow Schröder's or Dedekind's views of class while having no distinction between "falling under" and "subordination". As a matter of fact, from Frege's point of view, the aggregate view cannot coexist with the empty set:

> A class, in the sense in which we have so far used the word, consists of objects; it is an aggregate, a collective unity, of them; if so, it must vanish when these objects vanish. If we burn down all the trees of a wood, we thereby burn down the wood. There can be no empty class. (Geach and Black 1952, p. 89; see also p. 149)

But when Dedekind suggested that we rather exclude the empty class, as Gillies indicates, Dedekind must have known the difficulties which would result if he allowed the empty class without the distinction between "falling under" and "subordination":

> …we intend here for certain reasons wholly to exclude the empty system which contains no element at all, although for other investigations it may be appropriate to imagine such a system. (Dedekind 1888, pp. 45–46)

Unlike Dedekind, Schröder spoke of an empty class (Schröder 1890, 149; Geach and Black 1952, p. 150).

Frege claimed that this introduction of empty set yields insurmountable difficulties. Recent philosophers seem to have no qualms with Frege's accusation. Even when they want to interpret Schröder (or Cantor) as predecessors of the so-called iterative conception of set, they don't try to face Frege's attack directly. Rather they indicate that the iterative conception of set is not susceptible to Frege's condemnation:

> The faults with Schröder's and Cantor's conceptions are tied to their particular expressions and do not vitiate the modern iterative conception. (Resnik 1980, p. 206)

Then we should ask this: What is the difficulty Schröder's conception is doomed to invite?

If we have empty set ϕ, and if we don't distinguish between class membership and subset relation, we fall into a contradiction. Let A be a set which contains exactly n numbers. So $A = \{a_1, a_2, \ldots, a_n\}$. Now it is obvious that empty set ϕ is a subset of A. $\phi \subseteq A$. Since we don't distinguish between $\in$ and $\subseteq$, $\phi \in A$. So A comes to have one more member ϕ. Then A has $n + 1$ members. It is absurd (Gillies 1982, p. 10; Geach and Black 1952, p. 91).[12]

Schröder also knew this difficulty. So he imposed a restriction for the elements given as individuals within any given manifold. They should not contain within themselves any element of the some manifold as individuals (Geach and Black 1952, p. 91).

Interestingly enough, Church regarded this restriction as "a striking anticipation of the simple theory of types" (Church 1939, p. 408). According to him, the only difference lies in that "Schröder regards a unit class as identical with the single element which it contains". And he further indicates the facts that the paradoxes of set theory were unknown to Schröder and that Schröder needed the distinction of types as a substitute for the distinction between $\in$ and $\subseteq$.

Unfortunately, however, this stipulation was also criticized by Frege. Schröder defined 'identical' zero as follows: "0 is our name for a domain that stands to every domain or in the relation of being included". (Schröder 1890, 188; Geach and Black 1952, 94) Frege devoted a long argument for probing whether in this definition "the relation of being included" is taken as the sub($\subseteq$) or the subter($\in$) relation. His point is that, if we take the latter horn here. Then Schröder's stipulation cannot be satisfied. As long as we don't distinguish *sub* and *subter* relation, aggregate view of set, i.e., the conception of the class as consisting of individuals, cannot be upheld regardless of Schröder's stipulation.

> In the latter case, the conception leads to contradictions; in the former, we cannot abandon it if we are to make Schröder's stipulation at all realizable. (Geach and Black 1952, p. 97)[13]

3.2.3 The Cognate of the Distinction: Peano's $\in$ and $\supset$

In Sect. 2, we already mentioned that Peano had a distinction between $\in$ and $\supset$. Now it is time to press the question whether Peano's distinction is identical with Frege's distinction. In fact, Peano emphasized the importance of distinguishing class membership and class inclusion. His definitions of these relations clearly show the difference between them:

> The sign $\in$, which must not be confused with the sign $\supset$, applications of the inverse of logic, and a few other conventions, I have adopted so that I could express any proposition whatever. (Peano 1889, 102)

> The sign K means class, or aggregate of entities. The sign $\in$ means *is*. Thus a$\in$b is read a is a b; a$\in$k means a is a class; a$\in$p means a is a proposition. (Peano 1889, p. 107)

---

[12]Here I largely depend on Gillies' presentation. Frege's own reasoning is much more complicated.

[13]See Frege (1895, 97). If we mark the existence of the item as o and non-existence as x, then there are only different possibilities.

The sign $\supset$ means *is contained*. Thus a $\supset$ b means *the class a is contained in the class b*.

50. a, b$\in$k. $\supset$:.a$\supset$b:

=:x$\in$a. $\supset_x$.x$\in$b (Peano 1889, p. 108)

Frege himself conceded that peano also had clear distinction between class membership and class inclusion. In one of the footnotes to Frege (1895), he indicated that Peano used the signs '$\supset$' and '$\in$' instead of 'sub' and 'subter' (Geach and Black 1952, p. 94). Moreover, in an undated letter to Peano, Frege emphasized the importance of Peano's distinction. Unfortunately, the draft of this letter stops at a crucial point. However, from this incomplete draft, we can get some idea about how Frege might have evaluated Peano's distinction:

> It seems to me very important that you make a sharp distinction between singular and universal propositions by means of the signs $\in$ and $\supset$, since the relations of an object (individual) to a concept under which it falls and to the class to which it belongs are indeed quite different from the relation of a concept to a superordinate concept or of a class to a more comprehensive class. This difference is overlooked by many writers, as by Mr. Dedekind in his work, The Nature and Purpose of Numbers (Dedekind 1888).

> Its importance comes out especially when empty concepts and classes are taken into consideration. For these must be recognized, and you actually do this when you use the sign Λ. Of course, one must not then regard a class as constituted by the objects (individuals, entities) that belong to it, for in removing the objects one would then also be removing the class constituted by them. Instead one must regard the class as constituted by the characteristic marks, i.e. the properties which an object must have if it is to belong to it. It can then happen that these properties contradict one another, or that there occurs no object that combines them in itself. The class is then empty, but without being logically objectionable for that reason. Now from the singular proposition 'One is a fourth root of one' ($\vdash 1^4 = 1$) one can indeed infer: 'There are fourth roots of one'($\vert \ {}_{\top}{}^a{}_{\top}a^4 = 1$), but from the universal proposition 'All square roots of one are fourth roots of one'.... (Frege undated, Gabriel 1980, p. 109)

If Frege had thought that Peano fully understood the importance of the distinction, he would not have written a letter such as this. Frege reluctantly conceded that Peano also had a distinction between class membership and class inclusion. If one takes the aggregate view of sets, has no distinction—whether it be Frege's or peano's—and allows the empty class, then the difficulty which Schröder faced arises. If one takes the aggregate view, has no distinction, and allows no empty class, then like Dedekind one is devoid of empty class. Now, in Peano, we have the distinction and the empty class (Peano 1889, p. 107). But Peano seems to take the aggregate view (Ibid.). What would happen if we take such a position? It seems to me that Frege had no ready-made answer. Perhaps that might be the reason why in the letter he emphasized the fact that if we allow the empty set we should reject the aggregate view.

At this point, however, we should attention to the fact that the logical relationship between Frege's distinction and his argument against the aggregate view is not crystal clear. At one place Frege said that the lack of the distinction in Dedekind and Schröder is the result of their common mistaken view about class (Geach and Black 1952, 149). But he did not make it explicit by what process the aggregate view makes them blind to the distinction.

It is also interesting to note that in this letter Frege is saying as if there is no difference between the distinction between "falling under" and "subordination" and the distinction between "class membership" and 'class inclusion'. If there is no difference, why should we duplicate the same distinction? If there is a difference, why didn't Frege claim the superiority of his distinction? This indicates that Frege himself was probably not so clear about class and concept.

4   Why Does the Distinction Matter to Frege?

As we saw above, in his criticism of Dedekind and Schröder, Frege found them responding to the problem caused by ignoring his distinction between "falling under" and 'subordination". Further, he diagnosed the ultimate cause for their negligence of the distinction being the wrong views about classes they shared. That is nothing but the aggregate view of sets. In a word, Frege thought that the aggregate view of sets is incompatible with the empty set. Let me recapitulate a few important points. First, It is still not clear how Frege's distinction between "falling under" and "subordination" is related to the aggregate view of classes, for he did not elaborate why the aggregate view had made Dedekind and Schröder oblivious to the distinction. Secondly, Frege, in his letter to Peano and in his review of Schröder, did not sharply distinguish the distinction between "falling under" and "subordination" from the distinction between class membership and class inclusion.[14] However, the in-depth understanding of the significance of the former for Frege's work is possible only when they are sharply distinguished. This points to the possibility that Frege himself may not have fully understood the importance of his own distinction between "falling under" and "subordination". Finally, recent philosophers of mathematics do not take issue with Frege's criticism of the aggregate view of classes, even when they count Schröder or Cantor as the precursor of the so-called the iterative conception of sets. But why and how could the iterative conception of sets be immune to Frege's criticism of the aggregate view of classes? Is Peano's position, which takes the aggregate view of classes, is equipped with the distinction between class membership and class inclusion, and allows the empty set harmonious with the iterative conception of set?

---

[14]Cocchiarella disagrees with my opinion on the ground that these distinctions are equivalent "but that is because the distinction between membership and falling-under is one of the consequences of Frege's double-correlation thesis" (Cocchiarella 2015b).

4.1   In Search of Logical Objects

Above I raised the question "Why is the distinction between "falling under and "subordination" so important?". It seems hard to find precedents to try to answer this question. Even some attempted answers are not quite satisfactory. For example, according to Resnik, Frege suggested that without this distinction the aggregate view cannot explain empty set or the intransitivity of membership relation (Resnik 1980, p. 206). No doubt, these are important matters. However, we cannot be satisfied by answers like this, for we want to understand what positive roles and functions this distinction has in Frege's thought. From this point of view, we are seduced by an attractive answer, which places this distinction in Frege's lifelong enterprise, i.e., the Logicist programme. In order to say that arithmetic is reducible to logic, Frege was compelled to have classes as logical objects. Resnik aptly represents this point of view:

> The success of Frege's analysis depends upon his view of classes. For only if classes are logical objects and class theory is a branch of logic can it be claimed plausibly that he has shown that number theory or other parts of arithmetic is reducible to logic. (Resnik 1980, p. 204)

It seems difficult to claim that classes are logical objects as long as one takes the aggregate view. Why are classes that are merely the aggregates of elements not mathematical but logical objects? Since Frege needed logical objects, he had to establish the view that class is something derived, whereas concept is something primitive (see Resnik 1980, p. 207).

What then is the relation between the distinction between concept and class on the one hand and the distinction between "falling under and "subordination" on the other hand? Gillies gives us an explicit answer: "Frege uses Schroder's difficulties here as another argument for taking 'concept' as more basic than 'set'" (Gillies 1982, p. 55). Here Gillies is talking about Schröder's difficulties which caused by his daring introduction of empty class without employing the distinction between "falling under" and "subordination". So, according to this point of view, the logicist program needs concepts rather than classes, and the proof that concepts are more primitive than classes needs the distinction between "falling under" and "subordination.

However, Gillies is putting the cart before the horse here. If Gillies is right, then the distinction between "falling under" and "subordination" is the necessary condition of the distinction between classes and concepts. The fact of the matter is that the opposite is the case. We can counter Gillies' claim by quoting Frege's own words:

> So if we want to make clear to ourselves the distinction between the two relations, we must regard classes as extensions of concepts and make this the basis of our interpretation. (Geach and Black 1952, p. 94)

What is the difference between concept and class? Wang would think that these are essentially the same thing:

Frege uses concepts and relation as the foundation stones upon which to erect his structure; Dedekind uses classes and the relation of an element belonging to a class. Nowadays we would think that they employ essentially the same thing. (Wang 1957, 156)

But, as Wang indicates, not for Frege. And we have already seen how much emphasis was given by Frege to the superiority of concepts over classes.

As is well-known, this distinction between concept and object is the sharpest and the most basic one.

The concept is predicative.*

*It is, in fact, the reference of a grammatical predicate. (Geach and Black 1952, p. 43)

We may say in brief, taking 'subject' and 'predicate' in the linguistic sense: A concept is the reference of a predicate; an object is something that can never be the whole reference of a predicate, but can be the reference of a subject. (Geach and Black 1952, 47)

First of all, I must emphasize the radical difference between concepts and object, which is of such a nature that a concept can never substitute for an object or an object for a concept. (Kluge 1980, 4)

It is simply beyond the scope of this paper to discuss this important and much discussed distinction extensively. My point here is merely that the distinction between "falling under" and "subordination" depends upon the distinction between class and concept, and further upon the distinction between concept and object. It means that we can understand the distinction between "falling under" and "subordination" only when we scrutinize the distinction between class and concept and the distinction between concept and object, and that both distinctions are highly controversial. Thus it is understandable why the distinction between "falling under" and "subordination" has been given a relatively smaller spotlight. In this regard, however, we should not bypass the fact that the distinction between "falling under" and "subordination" seems to be the ultimate distinction aimed at by Frege.

4.2 Frege's Analysis of the Statements of Number

A much more plausible hypothesis is that this distinction is indispensable for Frege's analysis of the standard usage of the statements of number. As is well-known, Frege (1884) is meant to solve the problem "What Number is". The task was to settle the problem of the definability of Number:

Now here it is above all Number which has to be either defined or recognized as indefinable. This is the point which the present work is meant to settle. (Frege 1884, 5; also II7)

In order to define Number, Frege should analyze the general logical form of the statements of Number. At this stage, the distinction between "falling under" and "subordination" seems to be needed. The statements of Number have the form "there are nFs". This seems to be very similar to the form of the propositions such as "there are blind men". But, according to Frege, Numbers are not properties of objects. Rather, "the content of a statement of Number is an assertion about a concept" (Frege 1884, p. 59, 46, 57). Kremer, after having described this statement as one of the two basic principles of Frege (1884), aptly reveals its implication as follows:

> In making a statement of number of the form 'there are nF's, we convey information about the concept F: namely, *that n objects fall under it*". (emphasis is mine) (Kremer 1985, 313)

Also, we know that Frege avoided calling a number a property of a concept (Frege 1884, 68). Thus, to understand the difference between "falling within", "subordination", and "falling under" is a prerequisite to understand Frege's analysis of the statements of Number.

4.3   Frege's Function-Correlates

As is well-known, the distinction between function and object is the most fundamental distinction in Frege's ontology. This distinction corresponds to the distinction between function and argument in his conceptual notation which is meant to replace the traditional subject/predicate analysis of propositions. Of course, Frege borrowed the notion of function from mathematics, but he thought that the contemporary mathematicians had a very confused, if not inconsistent, idea about what a function is. Frege himself generalized the concept of function so that it may cover concepts as well as mathematical functions. For Frege, a concept is a function whose value is always a truth-value. According to him, the nature of functional expressions is essentially predicative. They are incomplete expressions with gaps which are to be filled by arguments. Corresponding to functional expressions, there are functions that are unsaturated entities. Object is nothing but something that is not a function. Thus, every entity in Frege's ontology is either a function or an object. Typical examples of Frege's objects are truth-values and value-ranges. Commentators of Frege also cite individuals, function-correlates, numbers, value-ranges, and truth-values.

What is remarkable is that values-ranges and concept-correlates are considered separately as Fregean objects. However, Cocchiarella's discussion of Frege makes it quite plausible that Frege's function-correlates are nominalized predicates, and that Frege identified function-correlates with value-ranges. To the best of my knowledge, Cocchiarella is the first, and possibly the only one, who explicitly claims that Frege equated value-ranges with concept-correlates. It is somewhat understandable why other commentators failed to notice this equation, for philosophers such as R. Wells believe that the notion of value-range is the obscurest of Frege's basic notions. But Wells says that it is very important because "it is his [Frege's] nearest approach to the notion of class". What Wells and others have in mind is the notion of class in the sense of being composed of its members. And, as long as one understands value-ranges as class in the sense of being composed of its members, there is no hope of seeing the equation. Cocchiarella flatly rejects this popular way of understanding value-ranges. According to him, they are "the saturated logical objects that Frege also informally called concept-correlates". That means, as Cocchiarella makes clear, that "value-ranges are for Frege the denotata of nominalized predicates".

Given the general distinction between function and object, if Frege came to have a unique category of function-correlate, i.e., the reference of nominalized predicates, how is it differentiated from function, and how is it related to it? In order to avoid confusion, Frege understood function as function itself, and contrasted it with function-correlate. Function itself is an incomplete entity, while function-correlate is an object. By using the lambda operator of Church, Cocchiarella expressed function itself as $[\lambda_x \varphi_{(x)}]()$, and function-correlate as $[\lambda_{x}(\varphi_x)]$. Perhaps Cocchiarella wants to present by these expressions Frege's metaphorical idea of unsaturatesness of functions. The empty parentheses in the former suggests the incompleteness of functional expression and the unsaturatedness of the corresponding function itself. For example, corresponding to the predicate '… is a horse', there is a function "… is a horse" designated by it. The nominalized form of the predicate is 'horseness', and its reference is the function-correlate "horseness".

4.4 The Distinction Between Falling Under and Subordination Versus the Distinction Between Class Membership and Class Inclusion

Then, how is Frege's ontology of function, function-correlate, and object related to his distinction between "falling under" and "subordination"? (See Park 1990, Angelelli 1979, 1984, Cocchiarella 1988, 2007, 2015a, Bergmann 1959, 1968) Let us synthesize all discussions above in order to deepen our understanding of the distinction. Now, the singular proposition "Socrates is mortal", which was simply understood as an example of "falling under", can be said to express unequal second-level relation, in which an object (Socrates) is subsumed under a first-level concept (- is mortal). Since Socrates and "- is mortal" belong to different levels, their relation is an unequal relation. Since "- is mortal", which plays the role of filling the empty part of the "falling under" relation itself, is itself a first-level function, the "falling under" relation itself becomes a second-level function. When symbolized, the fact that the "falling under" relation is an unequal second-level function can be expressed as follows:

$$(\exists R^2)\left(\left[\lambda_{xy}(\exists G)(x = G \& G_{(y))})\right] = R\right) \text{ (Cocchiarella 1987, p. 94)}$$

On the other hand, the universal proposition "All men are mortal", which was simply understood as an example of "subordination", can be interpreted as expressing the fact that an equal second-level relational concept, in which a first-level concept $[\lambda_x \text{ Man}_{(x)}]()$ is subordinated to another first-level function $[\lambda_x \text{ Man}_{(x)}]()$, is realized (Ibid. p. 77). Since the empty parts of the "subordination" relation itself are filled out by the first-level concepts that are themselves unsaturated, the "subordination" relation itself becomes an equal relation. Also, since both

$[\lambda_x \text{ Man}_{(x)}]()$ and $[\lambda_x \text{ Mortal}_{(x)}]()$ are first-level concepts, the "subordination" relation itself becomes an equal relation.

On the other hand, if the distinction between "falling under" and "subordination" is given, and further ontology of object and function is given, in accordance with what Cocchiarella calls "Frege's double correlation thesis" classes as concept-correlates can be gotten. According to Frege's double correlation thesis,

> all second-level concepts can be correlated one-to-one with certain special first-level concepts which in turn can be correlated one-to-one with special objects called concept-correlates. (Cocchiarella 1987, p. 158)

As Cocchiarella points out, Frege himself presented this thesis as follows:

> Second level functions can be represented in a certain manner by first level functions, whereby the functions that appear as arguments of the former are represented by their value-ranges. (Frege 1891, Geach and Black 1952, pp. 26–27)

In Cocchiarella's formulation, it can be expressed as follows:

$$(\forall Q)(\exists F)(\forall G) \left[(Q_x)G_x \leftrightarrow F(\varepsilon' \ G(\varepsilon)) \right]$$ (Cocchiarella 1987, p. 79, 158)

Here, the symbol $\acute{\varepsilon} \ G(\varepsilon))$, called spiritus lenis or smooth breathing operator, is to generate complex singular term representing value-ranges. Furthermore, as Cocchiarella claims,

> all concepts of whatever level can in effect be correlated with first level concepts, and these in turn can be correlated with their extensions. Frege's entire hierarchy of concepts, accordingly can be collapsed into the universe of first level concepts, which in turn can be correlated with the classes that are their extensions. (Cocchiarella 1987, p. 159)

If we have only the distinction between class membership and class inclusion, the singular proposition "Socrates is mortal" would amount to a relation between an object and another object. Also, the universal proposition "All men are mortal" would be a relation between two classes, and would thereby again become merely a relation between an object and another object. As a consequence, structural analysis in terms of "falling under" or "subordination" would be impossible. Even if it might be possible to explain how classes are formed by adopting the iterative conception of set, and thereby to secure a slightly strengthened position, it would be impossible to reveal the difference between singular and universal propositions, as was done by the distinction between "falling under" and "subordination".

References

Ackrill, J. L. (1957). Plato and the copula: Sophist 251–59. In G. Vlastos (Ed.), *Plato: A collection of critical essays, Vol. 1: Metaphysics and epistemology* (pp. 210–222). Garden City, NY: Anchor Books.

Ackrill, J. L. (1963). *Aristotle's categories and de interpretatione*, Translated with notes and glossaries. Oxford: Clarendon Press.

Angelelli, I. (1967). *Studies on Gottlob Frege and traditional philosophy*. Dordrecht-Holland: D. Reidel Publishing Company.

Angelelli, I. (1979). "Class as one" and "class as many". *Historia Mathematica, 6,* 305–309.

Angelelli, I. (1984). Frege and abstraction. *Philosophia Naturalis, 21*(2–4), 453–471.

Aristotle, *Categories* in Ackrill (1963).

Bergmann, G. (1959, 1968). Frege's hidden nominalism. In *Meaning and existence* (pp. 205–224). Madison: The University of Wisconsin Press.

Boolos, G. (1971). The iterative conception of set. *Journal of Philosophy, 68,* 215–232.

Church, A. (1939, 1979). Schröder's anticipation of the simple theory of types. *Erkenntnis, 10,* 407–411.

Cocchiarella, N. B. (1987). *Logical studies in early analytic philosophy*. Columbus: Ohio State University Press.

Cocchiarella, N. B. (1988). Predication versus membership in the distinction between logic as language and logic as calculus. *Synthese, 77,* 37–72.

Cocchiarella, N. B. (2007). *Formal ontology and conceptual realism*. Dordrecht: Springer.

Cocchiarella, N. B. (2015a). *On predication, a conceptualist view*. Forthcoming.

Cocchiarella, N. B. (2015b). "Letter to Park" (personal communication).

Copi, I. (1982). *Introduction to logic* (6th ed.). NY: Macmillan.

Dedekind, R. (1872). *Continuity and Irrational Numbers* in Dedekind (1963), pp. 1–27.

Dedekind, R. (1888). *The Nature and Meaning of Numbers* in Dedekind (1963), pp. 30–115.

Dedekind, R. (1963). *Essays on the theory of numbers* (W.W. Beman, Trans.). New York: Dover.

De Morgan, A. (1847). *Formal logic or the calculus of inference necessary and probable*. London: Taylor and Walton. (Elibron Classics Replica Edition published in 2005 by Adamant Media Corporation).

Frege, G. (undated). "Letter to Peano". In G. Gabriel et al. (Eds.), (1980)

Frege, G. (1884), *The foundations of arithmetic*, second revised edition. (J. L. Austin, Trans.) Evanston-Illinois: Northwestern University Press (1968).

Frege, G. (1891). Function and concept. *Geach and Black, 1952,* 21–41.

Frege, G. (1892). On concept and object. *Geach and Black, 1952,* 42–55.

Frege, G. (1893). *The basic laws of arithmetic* (M. Furth, Trans. & Ed.). Berkeley and Los Angeles: University of California Press (1964).

Frege, G. (1895). A critical elucidation of some points in Schröder's algebra der Logik. *Geach and Black, 1952,* 86–106.

Gabriel, G., et al. (Eds.). (1980). *Gottlob Frege: Philosophical and mathematical correspondence* (Ed.). Chicago: The University of Chicago Press.

Geach, P., & Black, M. (Trans. & Eds.). (1952). *Translations from the philosophical writings of Gottlob Frege* (2nd Ed.). Oxford: Basil Blackwell (1960).

Gillies, D. (1982). *Frege, Dedekind and Peano on the foundations of arithmetic*. Assen: Van Gorcum.

Hamblin, C. L. (1970). *Fallacies*. London: Methuen.

Hintikka, J. (1979). Frege's hidden semantics. *Revue Internationale de Philosophie, 130,* 716–722.

Hintikka, J. (1981). Semantics: A revolt against Frege. In G. Fløistad (Ed.), *Contemporary philosophy: A new survey* (Vol. 1, pp. 57–82). The Hague: Martinus Nijhoff.

Jourdain, P. E. B. (1912). Gottlob Frege. In G. Gabriel et al. (Eds.). (1980).

Kant, I. (1871, 1879). *Critique of pure reason*. Unabridged Edition. (N. K. Smith, Trans.). New York: St. Martin's Press (1965).

Kluge, E.-H. W. (1971). *Gottlob Frege: On the Foundations of Geometry and Formal Theories of Arithmetic*, translated with and Introduction by E.-H. W. Kluge, New Haven and London: Yale University Press.

Kluge, E.-H. W. (1980). *The metaphysics of Gottlob Frege: An essay in ontological reconstruction*. The Hage: Martinus Nijhoff.

Kneale, W., & Kneale, M. (1962). *Development of logic*. Oxford: Clarendon Press.

Kremer, M. (1985). Frege's theory of number and the distinction between function and object. *Philosophical Studies, 47,* 313–323.

Lambros, C. (1976). Are numbers properties of objects? *Philosophical Studies, 29*(1976), 381–389.

Park, W. (1990). Scotus, Frege, and Bergmann. *The Modern Schoolman, 67,* 259–273.

Park, W. (1992). Frege's distinction between 'falling under' and 'subordination'. *Yonsei Philosophy, 4,* 53–80. (in Korean).

Park, W. (1997). *In search of science lost: An introduction to philosophy for scientists.* Seoul: Damron. (in Korean).

Parsons, C. (1982). Objects and logic. *Monist, 65,* 491–516.

Peano, G. (1889). The principles of arithmetic, presented by a New Method. In H.C. Kennedy (Trans. & Ed.), *Selected works of Giuseppe Peano.* Toronto: University of Toronto Press (1973).

Porphyry. (1966). *Isagoge.* In Aristotelis Latinus (Vol. 1, pp. 6–7). Categoriarum, Supplementa, Desclèe de Brower, Brugels—Paris.

Porphyry. (1975). *Porphyry The Phoenician Isagoge.* Translation, introduction and notes by E. W. Warren. Toronto: The Pontifical Institute of Mediaeval Studies.

Prior, A. N. (1963). *Formal logic* (2nd ed.). Oxford: Clarendon Press.

Quine, W. V. (1940). *Mathematical logic.* Cambridge: Harvard Univ. Press.

Resnik, M. D. (1980). *Frege and the philosophy of mathematics.* Ithaca: Cornell University Press.

Russell, B. (1903). *Principles of Mathematics.* Cambridge: Cambridge University Press.

Russell, B. (1914). Our Knowledge of the External Worls, London: George Allen and Unwin.

Russell, B. (1959). *My philosophical development.* London: Unwin Books.

Shcröder, E. (1890). *Vorlesungen über die Algebra der Logik,* reprinted. Bronx: Chelsea Publishing Company (1966).

Sluga, H. (1980). *Gottlob Frege.* London: Routledge and Kegan Paul.

Wang, H. (1957). The axiomatization of arithmetic. *Journal of Symbolic Logic, 22,* 145–158.

William of Sherwood. (1966). *Introduction to logic.* Translated with an introduction and notes by N. Kretzmann. Minneapolis: University of Minnesota Press.

Chapter 3
Scotus, Frege, and Bergmann

Abstract If N. Cocchiarella's recent discussion of Frege's function-correlate is correct, we have reason to assimilate Frege's ontology to the Avicennian-Scotistic tripartite ontology of individuals, universals, and common natures in themselves. Further, to the extent that Scotus' ontology is similar to Frege's ontology, we may have indirect evidence concerning how Bergmann would think about such an interpretation of Scotus' *haecceitas* ontology. I want to show that current treatments of individuation can be seriously challenged by the possible return of the common nature. My strategy will be as follows. In Sect. 1, I shall discuss Cocchiarella's thesis regarding Frege's function-correlates. In Sect. 2, by using Cocchiarella's thesis, I shall try to compare Frege's ontology with the Avicennian-Scotistic tripartite ontology. Some of the similarities and differences between these two ontologies will become clearer in the process. In Sect. 3, I shall examine Bergmann's interpretation of Frege's ontology. After having drawn attention to how Bergmann criticizes Frege's introduction of concept-correlates and value-ranges, we may understand, in Sect. 4, how Bergmann would view Frege's and Scotus' tripartite ontologies. Hopefully the peculiarity of Bergmann's theory of universals and his bare particular theory of individuation will stand out clearly against the background of Frege's and Scotus' ontologies.

Keywords Bare particular · Bergmann · Cocchiarella · Frege · *Haecceitas* · Individuation · Ontology

Elsewhere, I pointed out some crucial differences between Scotus' *haecceitas* and Bergmann's bare particular, even though, as individuators, they share many interesting characteristics.[1] While Bergmann's bare particular is a monentary individual, *haecceitas* is a continuant. While Bergman's bare particular is an individual which supports or possesses characteristics, *haecceitas* is not an individual but a principle

---

This chapter was originally published as Park (1990b).

---

[1]Park (1990a). See also, Gracia (1982), pp. 221–222, for some of the similarities between *haecceitas* and the bare particular.

© Springer International Publishing AG, part of Springer Nature 2018
W. Park, *Philosophy's Loss of Logic to Mathematics*, Studies in Applied Philosophy, Epistemology and Rational Ethics 43, https://doi.org/10.1007/978-3-319-95147-8_3

of individuation. Further I went on to hint at the two structural roots from which such differences between *haecceitas* and the bare particular stem. First, Scotus and Bergmann philosophize under two different systems of logic and theories of predication. While Scotus is a pre-Fregean who does not distinguish between class inclusion and membership, Bergmann is a post-Fregean to whom the method of ideal language, i.e., the language of *Principia Mathemarica*, is indispensable for philosophizing. Second, they seem to address different problems of individuation. While there is no place for a common nature in Bergmann's ontology, Scotus' problem of individuation is inconceivable without the common nature in itself. In this chapter, I would like to discuss these two structural roots of the differences between *haecceitas* and the bare particular a bit further.

Interestingly, Frege looms large in understanding both of these structural differences between Scotus' and Bergmann's ontologies. Frege is the undisputed founder of modern logic. So, if the difference between Scotus' and Bergmann's ontologies are reducible to the difference between their logics, we have to ask first the implications of Frege's revolution in logic. Further, unlike Bergmann, Frege's ontology has a type of entity comparable to Scotus' common nature. For that reason, I believe that it is much easier to compare Scotus' ontology with Frege's ontology rather than to compare the former directly with Bergmann's ontology.[2] Indeed, if N. Cocchiarella's recent discussion of Frege's function-correlate is correct,[3] we have reason to assimilate Frege's ontology to the Avicennian-Scotistic tripartite ontology of individuals, universals, and common natures in themselves.[4] Further, since Bergmann himself discussed Frege's ontology quite extensively,[5] to the extent that Scotus' ontology is similar to Frege's ontology, we may have indirect evidence concerning how Bergmann would think about our interpretation of Scotus' *haecceitas* ontology. Some may doubt whether my approach would result any interesting historical or ontological insight. But, to those who have tried

---

[2]Some philosophers are unnecessarily hostile to the ontological reading of Frege. For example, M. Dummett treats the dispute between Kluge and Grossmann as "a beautiful example of the misconceptions of Frege that results from trying to fit him into the framework of traditional ontology." Dummett (1981), p. 177. In the same vein, he wrote: "Frege's ontology of objects, concepts. relations and functions is not a contribution to the traditional debate over particulars and universals, but clears it away and supersedes it." *ibid*, p. 169. It is hard to believe that such a remark is made by the author of a book one chapter of which is devoted to "Frege's Place in the History of Philosophy." Dummett (1973), pp. 665–684.

[3]N. B. Cocchiarella. "Frege. Russell, and Logicism: A Logical Reconstruction," in Cocchiarella (1987), pp. 64–118. A somewhat longer version of this paper appeared in L. Haaparanta and Hintikka (eds.), (1986), pp. 197–252.

[4]For an excellent introduction to Scotus' ontology, see J. Owens, "Common Nature: A Point of Comparison between Thomistic and Scotistic Metaphysics," in J. F. Ross (ed.), (1967), pp. 185–209. Also, see Park (1989), 188–192.

[5]G. Bergmann, "Frege's Hidden Nominalism" in Bergmann (1959, 1968), pp. 205–224; Bergmann, "Ontological Alternatives" in Bergmann (1964), pp. 124–157. Both papers are also found in E. D. Klemke (ed.), (1968), pp. 42–63 and pp. 113–156, respectively.

to fill the gap between medieval and contemporary theories of individuation, the implication of this paper will be obvious. Current treatments of individuation are being challenged by the possible return of the common nature.

My strategy will be as follows. In Sect. 1, I shall discuss Cocchiarella's thesis regarding Frege's function-correlates. In Sect. 2, by using Cocchiarella's thesis, I shall try to compare Frege's ontology with the Avicennian-Scotistic tripartite ontology. Some of the similarities and differences between these two ontologies will become clearer in the process. In Sect. 3, I shall examine Bergmann's interpretation of Frege's ontology. After having drawn attention to how Bergmann criticizes Frege's introduction of concept-correlates and value-ranges, we may understand, in Sect. 4, how Bergmann would view Frege's and Scotus' tripartite ontologies. Hopefully the peculiarity of Bergmann's theory of universals and his bare particular theory of individuation will standout clearly against the background of Frege's and Scotus' ontologies.

1   Cocchiarella's Thesis

Cocchiarella's recent discussion of Frege makes it quite plausible that Frege's function-correlates are nominalized predicates and that Frege identified function-correlates with value-ranges (Cocchiarella 1987, p. 76; See also Cocchiarella 1986). Indeed, Cocchiarella's discussion is the stepping stone for the comparison of Scotus and Frege in the subsequent sections of this chapter. But, in order to appreciate the significance of Cocchiarella's discussion, let me first briefly explain some of the elements of Frege's ontology.

As is well known, the distinction between function and object is the most fundamental distinction in Frege's ontology.[6] This distinction corresponds to the distinction between function and argument in his conceptual notation which is meant to replace the traditional subject/predicate analysis of propositions (Cf. Geach and Black (eds.), 1960, 1970, p. 2, 12–13; p. 31). Of course, Frege borrowed the notion of function from mathematics. But mathematicians in Frege's time did not have a clear idea about what a function is (Ibid., pp. 21–22). Frege generalized the concept of function in such a way that not only mathematical functions but also concepts can be subsumed under it. A concept is nothing but a function whose value is always a truth-value (Ibid., p. 30). According to Frege, the nature of functional expressions are essentially predicative (Ibid., p. 43). They are incomplete expressions with gaps which are to be filled by arguments. Corresponding to functional expressions, we

---

[6]For example, E. -H. W. Kluge wrote as follows: "Functions and objects—or more correctly, the notions of function and object—occupy a central place in Frege's metaphysics." [Kluge (1980), p. 41]. Frege himself discussed the distinction in his papers, "Function and Concept" and "On Concept and Object." Both are found in Geach and Black (eds.), (1960, 1970), pp. 21–41 and pp 42–55, respectively.

have functions that are unsaturated entities.[7] Now Frege characterized an object as something that is not a function.[8] Every entity in Frege's ontology is thus either a function or an object. As examples of objects. Frege did not cite ordinary individuals but truth-values and the so-called value-ranges (*Ibid.*, p. 32; Frege 1964, pp. 35–36). In commentators' list of Fregean objects, we find the following: individuals, function-correlates, numbers, value-ranges, and truth-values.[9]

As is clear from this list, both value-ranges and concept-correlates have been considered (separately) as Fregean objects (Bergmann 1959, 1968, p. 207). To the best of my knowledge, Cocchiarella is the first to claim explicitly that Frege equated value-ranges with concept-correlates.[10] Why couldn't all other commentators of Frege notice of that equation? The reason seems something like this. Many philosophers such as R. Wells believe that the notion of value-range is the obscurest of Frege's basic notions (Wells 1967, p. 13). But Wells says that it is very important because "it is his [Frege's] nearest approach to the modern notion of class" (Ibid.). What Wells and others have in mind is the notion of class in the sense of being composed of its members. And as long as one understands value-range as class in the sense of being composed of its members, there is no hope to see the equation. Cocchiarella flatly rejects this popular way of understanding of value-ranges.[11] According to him, they are "the saturated logical objects that Frege also informally called concept-correlates" (Ibid., p. 76). That means, as Cocchiarella makes clear, that "value-ranges are for Frege the denotata of nominalized predicates" (Ibid., p. 67 and p. 76). Now we turn to the comparison of Scotus' and Frege's ontology by assuming that Cocchiarella's interpretation is right.

2   Scotus and Frege

J. Owens admirably demonstrated that both Aquinas and Scotus developed their theories of common nature against the Avicennian background. Let us briefly review Owens' explanation of the situation. Owens summarized Avicenna's doctrine of the common nature in itself in terms of the following three theses: (1) a

---

[7]Cf. Frege, "Function and Concept," Ibid., p. 24: "I am concerned to show that the argument does not belong with the function to make up a complete whole: for the function by itself must be called incomplete, in need of supplementation, or 'unsaturated'."

[8]*Ibid.*, p. 32: "An object is anything that is not a function, so that an expression for it does not contain any empty place."

[9]For example, R. S. Wells, "Frege's Ontology" in Klemke (ed.), (1968), pp. 17–18; Bergmann (1959–1968), p. 207.

[10]Cocchiarella, *Op. cit.*, p. 76, C. Parson's paper "Objects and Logic" also contains invaluable discussion of the same issues. C. Parsons (1982), pp. 491–516, esp., pp. 498–505.

[11]Cocchiarella, *Op. cit.,* p. 76. Also, see his "The Double Correlation Thesis and Set Theories NF and ML" in Cocchiarella (1987). There he claims that the notion of a set as a class that is composed of its members is none other than the iterative concept of set.

common nature or essence is of itself neither singular nor universal; (2) the common nature in itself has no unity; (3) the common nature has its proper being. According to Owens, these Avicennian theses should be revised and qualified if the scholastic doctrine of the transcendentals is accepted. For if being and unity were thought to be coextensive, there can be nothing that has being but no unity. Aquinas gave up the Avicennian thesis (3), while Scotus denied thesis (2) by giving a real though less than numerical unity to the common nature in itself (Owens 1967).

Given such background information concerning the Avicennean-Scotistic tripartite ontologies, what would be the significance of Cocchiarella's thesis that value-ranges are concept-correlates? Though Cocchiarella himself values his own thesis in some different respects, what is important for my purpose in Cocchiarella's commentary is that in addition to the category of functions themselves, Frege had a unique category of concept correlates, i.e., the denotata of nominalized predicates. Further if a concept-correlate can be equated with a value-range, as claimed by Cocchiarella, it may be the most important type of Fregean objects that are not ordinary individuals. As Cocchiarella points out, in order to understand what Frege had in mind, we should be very careful not to confuse the function itself with function-correlates. In order to express the distinction by using Church's lambda operator, we have to use $[\lambda \times \varphi(x)]$ ( ) and $[\lambda \times \varphi(x)]$ for the denoting function itself and function-correlate respectively (Cocchiarella 1987, pp. 82–83). The empty parentheses in the former expression indicate that the functional expression is incomplete and the corresponding function itself is unsaturated. As an incomplete expression, the functional sign cannot be placed in the subject position in a sentence. But functions themselves may have properties of their own. In order to express such higher order predication, we need nominalization of functional signs. Thus the denotata of nominalized predicates are correlated to the functions themselves. They are saturated logical objects whose expression can be placed in the subject position of a sentence. For example, we have a predicate '—is horse,' and as its denotatum the function "—is horse" itself. The nominalized form of the predicate '—is horse' is the abstract singular term, 'horseness,' and its denotatum would be the functional-correlate "horseness". Then, there are three different ontological categories in Frege: individuals, function-correlates, and functions themselves. In Scotus, we have the common nature in itself, the common nature individuated in singulars, and the universal. Thus, we seem to have parallel tripartite ontologies in Scotus and Frege.

Our next step should be to see in what significant respect there are differences between Scotus' and Frege's tripartite ontologies. The most crucial point in this project must lie in the ontological status of functions and function-correlates: what kind of unity and being do Frege's functions and function-correlates have? In other words, we want to see Frege's position against the background of Avicenna's three theses and Scotus' corresponding theses.

Do functions and function-correlates have being? Yes! Since Frege treated function-correlates as objects, there is no doubt that function-correlates have being. Further, Frege seemed to claim that even functional expressions themselves have references (Dummett 1968, pp. 295–297). So, Frege's position seems to be similar to Scotus at least in that both allowed some kind of being to each of the terms in their tripartite ontologies.

If Frege's function-correlates and functions themselves have some kind of being, then our next question should be the following: Do they also have unity?; If so, what kind of unity and what kind of being? But we have to resolve several problems at this point, because it is by no means clear whether Frege used 'being' always in the same sense. L. Haaparanta's discussion of the equivocity of being in Frege seems to be relevant here. She writes:

> The above discussion shows that Frege assumes 'exists' and 'is' of existence to have two readings: they may refer either to an empty first-order concept, which he tries to treat from a pragmatic point of view, orto a meaningful second-order concept. In the former case the existential statement becomes meaningful if it is transformed into a metalinguistic statement which expresses that a given proper name has a reference. In the latter case the statement tells us that a concept is instantiated, i.e., that there is an object which has a given property. First-order existence is formalized by means of the existential quantifier and the symbol for identity, while second order existence is expressed by means of the existential quantifier and the symbol for predication (Haaparanta 1985, pp. 140–141).

Haaparanta philosophizes in the tradition of Hintikka, who has emphasized the fact that though many authors must have noticed the ambiguity of 'is' before Frege, it was Frege who for the first time clearly distinguished between these several meanings of 'is': (1) 'is' of identity, (2) 'is' of existence, (3) 'is' of predication, (4) 'is' of class inclusion.[12] Now, what Haaparanta points out in the above passage is this: Frege found an ambiguity even if one concentrated on 'is' of existence. 'Is' of existence may refer either to a first-order existence or to a second-order existence. On the ground that there are two such readings of 'is' of existence, she claims that Frege had an equivocity view of existence. But, if she is right, there seems to be an apparent conflict between Frege's and Scotus' views of being. For Scotus has been known as a champion of the univocity view of being.

But I believe that the alleged conflict between Scotus' and Frege's views of being is merely apparent. Let me explain why I believe so. Above all, Haaparanta's expression 'Frege's equivocity view of existence' seems to me a misnomer. Indeed, while Frege distinguished between the first-order concept of existence and the second-order concept of existence, Scotus was not armed with quantification theory, and consequently had the first-order concept of being only. But what seems relevant to our purpose here is the first-order concept of existence. Let us take an example of the second-order concept of existence: $(\exists x)$Human x. Haaparanta would say that in this case "a concept is instantiated. i.e., that there is an object which has a given property" (Haaparanta 1985, p. 141). But is she referring to "humanity" that is a function-correlate or "____is human" that is a function in itself? Does she interpret the formula as "there is an object which falls under humanity" or as "there is an object which is human"? When we raised the question as to whether the being of a function-correlate is the same as the corresponding function itself, the issue is

---

[12]J. Hintikka, "Semantics: A Revolt against Frege" in G. Floistad (ed.), (1981), p. 72. Also see papers in S. Knuuttila and J. Hintikka (eds.), (1986).

not whether one of them is a first-order concept while the other is a second-order concept. Rather, we wanted to know whether we can place function-correlates and functions in the place of "g" in the formula '(∃x) g = x.' In other words, if the first-order concept of existence is formalized, as Haaparanta claims, by means of the existential quantifier and the symbol for identity, we are raising the question as to whether a function or a function-correlate can be placed in the argument position. The answer is negative as to the functions in themselves. For function itself is unsaturated, and only saturated object can fill the empty place. So, even if one were to use the same sense of 'being,' we are not prejudging the issue concerning the difference between a function and a function-correlate with respect to being.

Now we have to note that Scotus' univocity view of being is one of the most obscure among his theories. As C. Shircel reported, there are several divergent approaches to his univocity view of being (Shircel 1942, pp. 3–4). I cannot discuss all these serious problems here. The only point I want to make is that even if Scotus would use the same sense of 'being' when he discussed the being of common nature in itself and the being of a universal, he could still allow different kinds of being to them. Even if Scotus used the term 'being' univocally to both God and creatures insofar as he was dealing with them as *ensinquantumens*, he granted different degrees of being to them. As was pointed out, Scotus did not know about the second-order concept of being. Further, it is also clear that Scotus would allow the being proper to a common nature in itself that has lesser than numerical unity because unity follows upon being. Then, the being of a universal and the being of the common nature in itself must be different, as the unity of universal is different from the unity of the common nature in itself. I would like to believe that Scotus used the first-order concept of existence in both cases, while fully conceding that the common nature in itself and the universal are beings indifferent degrees. That means, even though medieval philosophers had the first-order concept of existence only, the question still remains as to whether they would allow the same kind of being to the common nature in itself and to the universal. In a nutshell, I believe that we are making a category mistake if we detect a conflict between the so-called Frege's equivocity view of existence and the so-called Scotus' univocity view of being.

Now let us return to the problem of the unity and being. As was pointed out, Scotus acknowledged more than one type of unity (Park 1989). On the one hand, the common nature in itself is neither individual nor universal; it has a lesser than numerical unity. On the other hand, individuals have numerical unity. Finally, universals also have a certain unity, i.e., unity of universals (or universal unity). I. Angelelli emphasizes this point when he writes:

While the Aristotelian tradition, as well as Kant, had considered concepts as unsaturated entities (in a sense partially identical to Frege's Ungesättigtheit), the fact that concepts or universals are well-determined entities had also been stressed. Of course, the "individuality" of concepts were regarded as weaker than the individuality of real things, as a *unitas formalis* and not as a *unitas numerica*; still, it was a*unitas*. (Angelelli 1976, p. 157)

Angelelli goes on to point out that such a positive aspect of universals is preserved in Frege. Further, he explains that Frege "ought to stress the 'unity', the 'individuality' of concepts" (1) in order to solve the difficulties of the Euclidean definition of number, and (2) in order to reject Cantor's position on abstraction (Ibid.). Thus, the situation is not so different in Frege's ontology from Scotus' ontology, as far as individuals and universals are concerned.

How about function-correlates, which corresponds to common natures in themselves? Did Frege allow numerical unity to function-correlates? The answer seems to be positive. For Frege allowed nominalized predicates to occupy the subject position or argument position in general in sentences. Function-correlates are saturated objects to which laws of logic are applicable. If so, in an important respect the parallel between Frege's and Scotus' tripartite ontologies would break down. For the common nature in itself in Scotus' ontology is devoid of numerical unity. Also, since unity follows upon being, unlike Scotus who granted being proper to the common nature in itself, Frege would not allow any difference between the being of function $\sim$ correlates and that of functions themselves.

But how could Frege allow numerical unity to function-correlates? How could horseness be an individual which has numerical unity? After all, if he did not see any difference between ordinary individuals and function-correlates, why did he distinguish between "falling under" and "falling within"? It is strange that Frege did not subdivide the realm of objects into two kinds:ordinary individuals and the other types of objects. This line of thought reminds us of the fact that Frege characterized objects as that which are not functions. Why didn't he start from ordinary individuals and then characterized functions? It is true that in several places Frege used ordinary individuals as examples of objects. But it is still very strange that he did not give ordinary individuals as examples of objects immediately after he characterized functions and objects in his major writings (Geach and Black 1960, 1970, p. 32; Frege (1964), pp. 35–36).

If Frege had to draw a sharp line between ordinary individuals and function-correlates, function-correlates would lose numerical unity. And that would be detrimental to Frege's logicism, if it means that one cannot place function-correlates in argument positions in sentences, particularly, in identity sentences. Be that as it may, it seems clear that Frege had some real difficulties with handling unity of objects and identity sentences.

But what would happen if one takes Scotus' strategy by giving lesser than numerical unity to function correlates? Ordinary individuals have numerical unity, while logical objects have lesser than numerical unity.[13] The crucial issue is whether an entity having lesser than numerical unity can be placed in argument positions in sentences. If that is possible, there would be nothing to lose from Frege's part.[14]

But, of course, there are lots of difficulties involved in converting Frege into a Scotist in such a way. For example, there seems to be a crucial difference between Scotus' and Frege's views regarding the ontological priority among the three types

---

[13]Cocchiarella (1987), p. 80 wrote as follows: "That is, unlike sets whose existence or being is constituted by their members, concept-correlates, and therefore value-ranges, are "logical objects" whose determination is given by Frege's double correlation thesis...".

[14]This issue needs extensive discussion of the role of the context principle in Frege as a prerequisite.

of entities involved. In Scotus' ontology, 'common nature in itself' is a first intention notion. The common nature in itself is neither an individual nor a universal; it needs individuation to become an individual and it needs to be contracted by *haecceitas* to the individual. Thus, in a sense it is prior to individual. Since universals are the result of abstraction from individuals, they may have least priority among the three. On the other hand, in Frege's ontology, function-correlates are merely given by correlation to functions. In that sense, we may say that functions themselves are prior to function-correlates (Cocchiarella 1987, p. 80).

With all such difficulties, however, it is quite impressive that we have parallel tripartite ontologies in Scotus' and Frege's ontologies. So, we now goon to examine Bergmann's interpretation of Frege's ontology in order to surmise how Bergmann would respond to our interpretation of Scotus' ontology.

3   Bergmann and Frege

There would be no place for the common nature in itself in Bergmann's ontology. For, according to him, there is no character that is not exemplified in the world.[15] But, in the previous sections, we saw that to some extent we can assimilate Frege's ontology to Scotus's ontology. That means, if Frege's notion of concept-correlate is a counterpart of the common nature in itself, it would not be Welcomed by Bergmann. As a matter of fact, Bergmann claims that the need for concept-correlates is the most obvious intrinsic flaw in Frege's system (Bergmann 1959, 1968, p. 218). In this section, I would like to examine briefly Bergmann's interpretation of Frege's ontology. By observing how Bergmann criticizes Frege and how he distances himself from Frege, we may pave the way for better understanding of Bergmann's position.

In his paper, "Frege's Hidden Nominalism," Bergmann claims that the structure of Frege's ontology is nominalistic (Ibid., pp. 205–224). To show this, Bergmann first draws a distinction between two uses of the dichotomy of realism and nominalism: (1) the broad use, and (2) the strict use. In the broad use, a realist discerns many kinds of existents while a nominalist discerns only a few. In the strict use, a realist counts some characters as existents, while a nominalist counts no characters as existents (Ibid., p. 206). In this context, Bergmann uses 'existent' for 'what philosophers, speaking philosophically, assert to "exist" (Ibid., p. 205). According to Bergmann, Frege is an exuberant realist in the broad sense, but "in the strict sense at least implicitly a nominalist" (Ibid., p. 207). Further, he claims that Frege's exuberant realism in the broad sense was forced by his nominalism in the strict sense (Ibid.). Furthermore, he claims that "the most obvious intrinsic flaw" of Frege's system is another consequence of Frege's hidden nominalism (Ibid.).

What does Bergmann mean by Frege's hidden nominalism? From Bergmann's point of view, "while within his system at least Frege succeeded in securing full

---

[15]It is what Bergmann calls the "principle of exemplification."

ontological status for his odd objects, he did not so succeed, even within the system, in the case of functions" (Ibid., p. 212). And he traces it back to what he calls "its root," i.e., "the contrast between exemplification and mapping" (Ibid.). In other words, Bergmann believes that by contrasting his exemplification ontology with Frege's function ontology, he can uncover the hidden nominalism of Frege.

Then, what is the crucial difference between Bergmann's exemplification ontology and Frege's function ontology? According to Bergmann, a realist starts from individuals and their characters, and uses 'Peter is blond' as the paradigm (Ibid., p. 208). That suggests two major claims. First, in this paradigm, both 'Peter' and 'blond' stand for existents (Ibid.). Second, the state of affairs that 'Peter is blond' stands for is actualized if the individual and the character enter into a certain relation—or what Bergmann prefers to say "nexus" [Ibid.]. And this nexus is nothing but exemplification (Ibid.). These two major claims imply, Bergmann pointsout, two further claims. First, the nexus is asymmetrical. While an individual may or may not exemplify a character, it is nonsense to say that a character exemplifies an individual. Second, in the paradigm 'Peter is blond,' the verb, i.e., 'is' of predication, stands for exemplification.

There may be many reasons why a realist adopts his position. But Bergmann wants to point out two weighty reasons why "in *some respect* individuals and characters are alike" (Ibid., p. 209). The first fundamental likeness is the principle of exemplification we saw above: (Supra, n. 16) "Just as there is no individual that is not qualified, so there is no character that is not exemplified" (Bergmann 1959, 1968, p. 209). The second likeness is that "neither an individual nor a character is the kind of entity...a sentence stands for" (Bergmann 1959, 1968, p. 209). According to Bergmann, this is another way of saying that 'Peter,' 'blond'and 'Peter is blond' stand for three different entities (Ibid.). Again, Bergmann pointsout, this second likeness may be expressed, in Frege's terminology, by saying "individuals and characters are equally unsaturated" (Ibid.).

On the other hand, Bergmann observes that Frege's ontology starts from numbers and their functions and uses 'x^2' as the paradigm (Ibid., p. 210). Bergmann's strategy in contrasting Frege's function ontology with his exemplification ontology is to show that "numbers and their functions differ from each other in the two fundamental respects in which...individuals and characters are alike" (Ibid., p. 211). If the relation between numbers and their function is so different from the relation between individuals and characters, Frege's strategy to apply his fundamental function/object analysis to a fact consisting of individual and character is a total failure.

Bergmann first wants to fix the meaning of function: "The current mathematical name for the crucial idea is *mapping*."[16] I cannot examine here whether his claim is correct and justified. What is important at this stage is to see why the notion of mapping does suit Bergmann's purposes. Apparently, Bergmann wants to emphasize the fact that the ontological status of functions is not as secure as that of

---

[16]*Ibid.,* p. 210. Interestingly, P. Tichy suggested an interpretation which directly contradicts this: "Frege tells us quite clearly that they [functions] are not mappings." Tichy (1988), p. 21.

objects. For he hastens to point out that "a function is a mapping rule, mapping each member of one of the classes upon one (and, in the paradigmatic case, only one) member of the other" (Bergmann 1959, 1968, p. 210). The heart of the matter, he believes, is that "a mapping rule…is a thing much more shadowy, much lessreal, less palpable, less substantial than the things mapped and mapped upon" (Ibid.).

After presenting functions as mapping rules, Bergmann points out the two fundamental respects in which numbers and their functions differ from each other while individuals and characters are alike. First, Bergmann believes that there is counterpart of the principle of exemplification in function ontology. For "…there are of course numbers whether or not they be either arguments or values of functions" (Ibid., p. 211). Second, while the notions of individuals and characters are equally saturated or unsaturated, "the notion of number neither contains nor presupposes that of a function. The latter, however, contains and presupposes that of the two ranges (of numbers)" (Ibid.).

As we have seen, Bergmann believes that ordinary individuals are one kind of object in Frege's system (Ibid., p. 206). Further, he enumerates the following entities as what would be counted as objects: "individuals, numbers, truth values, value ranges (class of objects), senses, propositions (thoughts), concept correlates" (Ibid., p. 207). If Frege had used 'existent' as Bergmann does, Frege would have agreed, Bergrnann assumes, that every Fregean object is an existent" (Ibid., p. 206) ln other words, Bergmann refuses to view Fregean objects, which seem odd to him, as existents. Bergmann declares that he rejects Frege's exaggerated realism (i.e., realism in the broad sense) (Ibid., p. 207).

As is clear from Bergmann's list of Fregean objects, Bergmann counts value ranges and concept-correlates as two different kinds of objects. Both kinds are odd to Bergmann. And, interestingly enough, he tries to show how Frege was forced to have each of these odd kinds of objects by the very logic of his nominalism. He does it for value-ranges in Sect. 3 of "Frege's Hidden Nominalisrn". Also, he does it for concept-correlates—the need of which is believed by Bergmann to be the most obvious intrinsic flaw of Frege's system—in Sect. 4 of the same paper. Let us take a look at each of these cases.

According to Bergmann, Frege was forced to introduce "classes (extensions, value ranges)" into his ontology by the very logic of his nominalism. (Ibid., p. 214) He writes:

> Corresponding to each concept, which according to him is not an object and (if I am right) at least implicitly not an existent, Frege "creates" another entity, which according to him is an object, namely, the class of all objects which, as he says, fall under the concept (Ibid.).

Then, he promises to give two reasons why Frege's nominalism forces him to create such an entity (Ibid.). The first reason is that "Frege cannot and in fact does not, specify conditions of identity for concepts and functions" (Ibid., p. 215). The problem arises because if function is a mapping rule, the denotata of the functional expressions 'x^2' and '$x^2 - x + x$' cannot be the same. But, Bergmann points out, Frege must have wanted to preserve "what as a mathematician he surely wanted to preserve, namely, the equation '$x^2 = x^2 - x + x$' and the truism for which it stands" (Ibid.). So, Bergmann believes that Frege's way out is "to interpret it as an identity not between the two functions but, rather, between extensions" (Ibid.).

The need for concept correlates is, Bergmann claims, again a consequence of Frege's hidden nominalism (Ibid., p. 218). He explains how such a need arises by referring to the infamous puzzle "The concept horse is not a concept". By taking sides with many commentators of Frege, he claims that it is a simple truism to say 'the concept horse is a concept' (Ibid.). But, in Frege's system, every saturated expression beginning with 'the' is a name which denotes an object. However, since a concept is not an object, according to Frege, 'The concept horse is a concept' is false. So, Bergmann explains that in order to avoid this nonsensical consequence, Frege introduced a new kind of object, i.e., the concept-correlate (Ibid.).

Bergmann goes on to point out that "once the new kind, the concept-correlate, has been introduced, one cannot escape answering the question in what relation, or connection…it stands to the other two, the concept itself and its extension" (Ibid.). But, according to him, there is no answer to the question [Ibid.]. That is the major reason why Bergmann calls the need for creation of concept-correlates as 'the most obvious intrinsic flaw of the system *as it stands*" (Ibid.).

Further, Bergmann believes that, unlike value-ranges, a very simple emendation eliminates the need for the reification of concept correlates as a further kind of objects. Bergmann proposes to consider Frege's three basic grammatical categories exemplified by the following: (α) 'Peter,' (β) 'Peter is blond,' (γ) 'blond' (Ibid., p. 219). Interestingly, Bergmann believes that only in the first two expressions do we have both a reference and a sense (Ibid.). "The pattern does not apply to the unsaturated expression γ" (Ibid.). That means, Bergmann thinks that there is no such thing as the reference of the concept itself in Frege. As it turns out from Frege's posthumous writings, however. Frege had the reference of concept itself in his system (Dummett 1968). By failing to note this, Bergmann thinks that "that dangling third entity, the concept correlate" is the trouble maker (Bergmann 1959, 1968, p. 219).

4   Scotus and Bergmann

We opened this chapter with discussion of the two possible structural roots of the differences between Scotus' *haecceitas* ontology and Bergmann's bare particular ontology: (1) Scotus was a pre-Fregean, while Bergmann was equipped with modern symbolic logic; (2) Scotus' problem of individuation is inconceivable without the common nature, while there is no place for a common nature in Bergmann's ontology. Since Frege looms large in both points not only as a founder of modern logic but also as having a tripartite ontology comparable to Scotus. We compared Scotus and Frege in terms of the being and the unity of the terms involved in both. Also, we examined Bergmann's interpretation of Frege's ontology. Thus, now we should be in a better position to compare efficiently Bergmann's ontology with Scotus' ontology.

What is most impressive is that Bergmann flatly denies the need for exuberant tripartite ontologies. As he criticizes Frege's function-correlates and value-ranges, he would also criticize Scotus' common nature in itself. As he criticizes Frege's functions as nominalistic, he would also criticize Scotus' universals in the same vein. Then, among the so-called triple status of universals, the only aspect left is *universale in re*. Indeed, Bergmann believes that his ontology is the truly realistic one in that it secures the place for universals in the physical world and not in a Platonic heaven or in some mind. Since there is no common nature in itself in his ontology, for him the problem of individuation cannot be that of "what makes the common nature in itself indivisible (orincommunicable)?" or "what makes the common nature in itself contracted to this or that individual?", as it was for Scotus. Thus, while the problem of individuation was primarily a problem of incommunicability and only derivatively a problem of distinction (or numerical difference) for Scotus, (Park 1988) for Bergmann it was just the problem of explaining the numerical difference between the individuals of the same kind. While universals are used for explaining the sameness of the individuals within the same kind in Bergmann's ontology, bare particulars are meant to explain the numerical difference of individuals within the same kind.

There seems to be another interesting point that enables us to uncover the differences between Scotus', Frege's, and Bergmann's ontologies. Bergmann correctly observes that Frege would view function-correlates as existents (in Bergmann's sense). But, from Bergmann's point of view, it is odd to view them as existents. In order for entities to be existents in Bergmann's sense, they should have either numerical unity as individuals or universality as universals (in Bergmann's sense). Now, by using Bergmann's criticism of Frege as a springboard, Scotus would make the following comments. Bergmann is right to the extent that he views, unlike Frege, function-correlates as neither individuals having numerical unity nor universals having universality. But Bergmann is wrong to the extent that he refuses to allow any unity, e.g., lesser than numerical unity, and thereby any corresponding being to function-correlates. Frege is right to the extent that he grants being and unity to the third type of entities, i.e., function-correlates. But Frege is wrong to the extent that he gives *numerical* unity to function-correlates.

These results do not seem to bring about, of course, the clean explanation of how the two structural differences between Scotus' and Bergmann's ontologies led to the differences between *haecceitas* and the bare particular. But our discussion may at least shed some light on the second structural difference, i.e., that Scotus and Bergmann seem to discuss different problems of individuation. By assimilating Frege to Scotus, I hinted at how to welcome the long awaited and inevitable return of the common nature. I believe that the current discussions of individuation without common nature are empty. As far as the first structural difference is concerned, I have to concede that not much was achieved in this paper. The emergence of modern logic must be at the core of the philosophico-historical explanation of the differences between Scotus and Bergmann. But, as it turns out, it is not easy to determine whether Bergmann's ontology, in particular his theory of individuation, is the background ontology of Frege's logic. For we have seen above that in many respects Frege is more similar to Scotus than Bergmann. Probably, we will be able to learn a lot by further comparison of Frege and Scotus. For example, one might want to exploit Frege's innovation in understanding Scotus' obscure theory of *haecceitas* or to modify Scotus' theory in some respects in the light of Fregean logic. On the other hand, one may use Scotus as a measure of Frege's ontological development. Frege had started his career as a mathematician, and worked on the borderline between mathematics and philosophy. By having a glimmering of function-correlates, wasn't Frege becoming a *lesser-than-subtle* doctor?

References

Angelelli, I. (1976). *Studies on Gottlob Frege and traditional philosophy*. Dordrecht: D. Reidel.

Bergmann, G. (1959, 1968). Frege's hidden nominalism. In *Meaning and existence* (pp. 205–224). Madison: The University of Wisconsin Press. See also Klemke (Ed.). (1968). *Essays on Frege* (pp. 42–63). Urbana: University of Illinois Press.

Bergmann, G. (1964). Ontological alternatives. In *Logic and reality* (pp. 124–157). Madison: The University of Wisconsin Press. See also Klemke (Ed.). (1968). pp. 113–156.

Cocchiarella, N. B. (1986). Frege. Russell, and logicism: A logical reconstruction. In Cocchiarella (1987), pp. 64–118. A somewhat longer version of this paper appeared in Haaparanta and Hintikka (Eds.), (1986). *Frege synthesized* (pp. 197–252). Dordrecht, Holland: D. Reidel.

Cocchiarella, N. B. (1987). *Logical studies in early analytic philosophy*. Columbus: Ohio State University Press.

Dummett, M. (1968). Note: Frege on functions. In Klemke (Ed.). *Essays on Frege* (pp. 295–297). Urbana: University of Illinois Press.

Dummett, M. (1973). *Frege: Philosophy of language*. New York: Harper and Row.

Dummett, M. (1981). *The interpretation of Frege's philosophy*. Cambridge: Harvard University Press.

Frege, G. (1964). *The basic laws of arithmetic, translated and edited with an introduction by M. Furth*. Berkeley and Los Angeles: University of California Press.

Geach, P., & Black, M. (Eds.). (1960 1970). *Translations from the philosophical writings of Gottlob Frege* (2nd ed.). Oxford: Basil Blackwell.

Gracia, J. J. E. (1982). *Suarez on individuation*. Milwaukee: Marquette University Press.

Haaparanta, L. (1985). *Frege's doctrine of being*. Helsinki: Acta Philosophica Fennica.

Haaparanta, L., & Hintikka, J. (Eds.). (1986). *Frege synthesized*. Dordrecht, Holland: D. Reidel.

Hintikka, J. (1981). Semantics: A revolt against Frege. In G. Floistad (Ed.), *Contemporary philosophy: A new survey*. The Hague: MartinusNijhoff.

Klemke, E. D. (Ed.). (1968). *Essays on Frege*. Urbana: University of Illinois Press.

Kluge, E.-H. W. (1980). *The metaphysics of Gottlob Frege: An essay in ontological reconstruction*. The Hague: Martinus Nijhoff.

Knuuttila, S., & Hintikka, J. (Eds.). (1986). *The logic of being*. Dordrecht: D. Reidel.

Owens, J. (1967). Common nature: A point of comparison between Thomistic and Scotistic metaphysics. In J. F. Ross (Ed.), *Inquiries into medieval philosophy: A collection in honor of Francis P. Clarke* (pp. 185–209). Westport: Greenwood Pub. Co.

Park, W. (1988). The problem of individuation for Scotus: A principle of indivisibility or a principle of distinction? *Franciscan Studies, 48,* 105–123.

Park, W. (1989). Common nature and Haecceitas. *FranziskanischeStudien, 71,* 188–192.

Park, W. (1990a). *Haecceitas* and the bare particular. *The Review of Metaphysics, XLIV,* 375–398.

Park, W. (1990b). Scotus, Frege, and Bergmann. *Modern Schoolman, 67,* 259–273.

Parsons, C. (1982). Objects and logic. *Monist, 65*(1982), 491–516.

Shircel, C. L. (1942). *The univocity of the concept of being in the philosophy of John Duns Scotus.* Washington, D.C.: The Catholic University of America Press.

Tichy, P. (1988). *The foundations of Frege's logic.* Berlin and New York: Walter de Gruyter.

Wells, R. S. (1967). Frege's ontology. In Klemke (Ed.), *Essays on Frege* (pp. 3–41). Urbana: University of Illinois Press.

Chapter 4
On Cocchiarella's Retroactive Theory of Reference

Abstract I attempt to get further insights from Cocchiarella's history and philosophy of logic in understanding the contrast of Aristotelian and Fregean logic. Recently Cocchiarella proposed a conceptual theory of the referential and predicable concepts used in basic speech and mental acts (Cocchiarella in Synthese 114:169–202, 1998). This theory is interesting in itself in that singular and general, complex and simple, and pronominal and nonpronominal, referential concepts are claimed to be given a uniform account. Further, as a fundamental goal of this theory is to generate logical forms that represent the cognitive structure of our speech and mental acts, as well as logical forms that represent only the truth conditions of those acts, it is an indispensable part of Cocchiarella's conceptual realism as a formal ontology for general framework of knowledge representation. In view of the recent surge of interest in his formal ontology by cognitive scientists and AI people, at least, Cocchiarella's theory of reference deserves careful examination. Above all, however, the utmost value of Cocchiarella's theory of reference must be found in its challenge against what he calls "the paradigm of reducing general reference to singular reference of logically proper names" that pervades the 20th century (Cocchiarella in Synthese 114:169–202, 1998, 170). The aim of the present chapter is to provide an impressionistic sketch of Cocchiarella's challenge.

Keywords Aristotle · Cocchiarella · Frege · Predication · Reference

1 Introduction

Recently Cocchiarella proposed a conceptual theory of the referential and predicable concepts used in basic speech and mental acts (Cocchiarella 1998).[1] This theory is interesting in itself in that singular and general, complex and simple, and

---

This chapter was originally published in Park (2001, pp. 79–89).

---

[1]It will be an interesting and meaningful project to probe exactly when and where Cocchiarella figured out the substantial part of his theory of reference.

© Springer International Publishing AG, part of Springer Nature 2018   47
W. Park, *Philosophy's Loss of Logic to Mathematics*, Studies in Applied Philosophy,
Epistemology and Rational Ethics 43, https://doi.org/10.1007/978-3-319-95147-8_4

pronominal and nonpronominal, referential concepts are claimed to be given a uniform account. Further, as a fundamental goal of this theory is to generate logical forms that represent the cognitive structure of our speech and mental acts, as well as logical forms that represent only the truth conditions of those acts, it is an indispensable part of Cocchiarella's conceptual realism as a formal ontology for general framework of knowledge representation.[2] In view of the recent surge of interest in his formal ontology by cognitive scientists and AI people, at least, Cocchiarella's theory of reference deserves careful examination.[3] Above all, however, the utmost value of Cocchiarella's theory of reference must be found in its challenge against what he calls "the paradigm of reducing general reference to singular reference of logically proper names" that pervades the 20th century (Cocchiarella 1998, 170). The aim of the present study is to provide an impressionistic sketch of Cocchiarella's challenge.

No one would deny the correctness of Cocchiarella's observation that "theories of reference in the 20th century have been almost exclusively theories of singular reference—i.e., theories of the use of proper names and definite descriptions to refer to singular objects" (Ibid., 169). Ironically, not many are qualified enough to evaluate his further note that this is in marked contrast with medieval theories and the traditional theories of reference that are part of syllogistic logic. For, to most of us brainwashed by the theory of reference in the 20th century, it is rather shocking that there were theories of general as well as singular reference. If Cocchiarella is right about the history of the theories of reference, his challenge against the current paradigm is retroactive as well as revolutionary.

According to Cocchiarella, even Russell in his 1903 Principles of Mathematics held such a theory in which quantifier phrases stand for denoting concepts. Only in his 1905 "On Denoting" Russell abandoned the earlier theory by counting not only definite descriptions but also all quantifier phrases as incomplete symbols. Cocchiarella also notes that the early Russell held a view not unlike the so-called new theory of reference in that he treated proper names as not having any meaning but merely indicating. Later, Russell distinguished between ordinary proper names, which are eliminable in terms of definite descriptions, and logically proper names, each of which "applies directly to just one object, and does not in any way describe the object to which it applies". Cocchiarella finds in Russell's logical atomism, in which logically proper names culminate, the clearest paradigm of the idea of eliminating all forms of general reference (Cocchiarella 1998, pp. 169–170).

Now Cocchiarella's choice of the term "paradigm" must be judiciously intended in view of his another shocking observation that "aside from this paradigm, there were no explicit arguments against theories of general reference in favor of singular reference" (Ibid., 170). The significance of this observation can become apparent, if

---

[2]Cocchiarella's early papers are found in Cocchiarella (1986), and Cocchiarella (1987). For his application of formal ontology to the problems of knowledge representation, see Cocchiarella (1995).

[3]We may cite Nicola Guarino and his formal ontology research lab in Padova, Italy as an example.

Cocchiarella is right in his claim that "in many ways, and however unwittingly, it is this paradigm for reducing general reference to singular reference that has sustained the so-called "new" theory of direct reference" (Ibid.). Interestingly enough, since some such impossibility arguments for general reference have been developed long later by Peter Geach in his provoking 1962 book *Reference and Generality*, Cocchiarella is anxious to show that Geach's arguments fail to apply to his own theory of general as well as singular reference in terms of the framework of conceptual realism. It is understandable that Cocchiarella is intrigued by such a philosophical polemic, for he is primarily a philosopher rather than a historian of philosophy in the 20th century. Though it must be both an indispensable and invaluable project to determine whether Cocchiarella demonstrates successfully the adequacy of his theory of reference by rebutting Geach's arguments, I decline to plunge into it in this chapter. For I believe that Cocchiarella's identification of the alleged paradigm must be an excellent historical conjecture regardless of the final result of the controversy between Geach and Cocchiarella. In what follows, I will attempt to make Cocchiarella's basic ideas more accessible to me and general public with a hope of fathoming both the origin of the paradigm and the origin of Cocchiarella's revolt against it.

2 Predicable Concepts and Nominalized Predicates

Probably, the first step in developing Cocchiarella's conceptual realism[4] is the introduction of "a syntactical operation that transforms predicates (and formulas) into abstract singular terms, i.e., into nominalized predicates (and formulas). That is transparent in his presentation of the core of conceptual intentional realism:

> The core of the theory, which we will only briefly describe here, amounts to an extension of standard second-order logic in which sentence forms and predicate expressions can be nominalized and allowed to occur as abstract singular terms on a par with individual variables – with the one modification that the first-order part of the logic is free of existential presuppositions for singular terms. Complex predicates are formed by means of λ-abstraction, so that where φ is a formula and $y_1 .... y_n$ are pairwise distinct individual variables, $[\lambda y_1....y_n]$ is an n-place predicate expression. A predicate expression is always accompanied by a pair of parentheses (and commas if it is relational) when it occurs in its functional role as a predicate, as in $F(x_1....x_n)$ and $[\lambda y_1....y_n\,'\varphi](x_1....x_n)$, where F and $[\lambda y_1....y_n\,'\varphi](x_1....x_n)$ are n-place predicate expressions....To nominalize a predicate expression, we simply drop the parentheses (and commas) and allow the result to occur as an abstract singular term – as in G(F), $G([\lambda y_1...y_n\,'\varphi])$, R(x, F) and even G(G) and R(x, R)

[4]According to Cocchiarella, there are two kinds of realism in conceptual realism. One is an intensional realism which concerns the denotata of nominalized predicates and propositional forms. The other is a natural realism which has to do with natural kinds and natural properties. I briefly discussed the former in Park (1990) and Park (1992), The latter was also presented in Park (1996).

where G is a 1-place, R a 2-place, and F and $[\lambda y_1 \dots y_n \varphi]$ are n-place predicates (for $n \in \omega$). (Cocchiarella 1998, 171)

Cocchiarella emphasizes the importance of distinguishing predicable concepts, which are values of the predicate variables, from the abstract objects that are their intensional contents. The latter are denoted by nominalized predicates as abstract singular terms. These predicable concepts are, according to Cocchiarella, "intersubjectively realizable (and in that sense *objective*) cognitive capacities, or cognitive structure based upon such capacities, to characterize and relate objects in various ways (Cocchiarella 1998, 172).[5] In particular, Cocchiarella takes them to be "the cognitive capacities that underlie our rule-following abilities in the correct use of predicate expressions" (Ibid.).[6]

Here Cocchiarella makes it explicit that he assumes in this regard that predicable concepts are those features of thought and communication that determine the truth conditions of thought and communication. Through the exercise of predicable concepts as these cognitive capacities, according to Cocchiarella, our speech as well as mental acts are informed with a predicable nature. As concepts, predicable concepts are not objects, and thereby cannot be values of individual variables, but are "rather unsaturated cognitive structures that are realized (saturated) in particular mental acts—including speech acts as overt forms of mental acts having a communicative role" (Cocchiarella 1995, 699). Here the epithet "unsaturated" evidently indicates the similarity of Cocchiarella's notion of unsaturated nature of predicable concepts and Frege's notion of unsaturated nature of functions,[7] though Cocchiarella claims that his notion of unsaturatedness is not the same as that of Frege's (Cocchiarella 1998, 172).

Cocchiarella emphatically points out that insofar as a nominalized predicate denotes anything at all, as a consequence, what it denotes cannot be concept that the predicate stands for in its role as a predicate. If so, what is the ontological status of what a nominalized predicate denotes? Cocchiarella writes:

That it denotes anything at all is a posit that is made in conceptual realism for most (but not all) concepts; and in particular the posit is that as an abstract singular term a nominalized predicate denotes an intensional object – specifically, the intensional content of the concept that the predicate stands for in its role as a predicate. Here, by the intensional content of a concept we mean a hypostatization, reification, or projection into the domain of objects of the truth conditions determined by the different possible applications or exercises of that concept as a cognitive capacity. It is by mean of such a projection, or conceptual nominalization, that we purport to denote the intensional content of a concept – which we also

---

[5]In Cocchiarella (1995), 699, we find one further function of such cognitive capacities, i.e., "to identify objects".

[6]Cocchiarella (1995, in 699), characterized concepts as "intersubjectively realizable (and in that sense *objective*) cognitive structures that underlie our ability to think and communicate with one another". Note the special role of the cognitive capacities that ground predicable concepts.

[7]See, in particular, Cocchiarella's "Frege, Russell, and Logicism" in Cocchiarella (1987). Among Frege's writings, "Function and Concept", and "On Concept and Object" must be relevant in this connection. Both papers are found in Frege (1952). See also Park (1990).

call the *intension* of the concept – as if it were an independently existing real Platonic form. (Cocchiarella 1998, 172–173)

The ontological position portrayed in these sentences seems to be pregnant with many interesting controversial contents. We will be able to touch upon some aspects of it as we go along, though we should hurry to move to the heart of our present concern, i.e., referential concepts as the backbone of Cocchiarella's theory of general as well as singular reference.

3   Referential Concepts

3.1   General Reference

Cocchiarella introduces now referential concepts and their abstract intensional contents in addition to predicable concepts and their abstract contents (Ibid. 173). Referential concepts are, according to him, "cognitive capacities that have a structure complementary to predicable concepts the way that quantifier phrases are complementary to predicate expressions (or the way that noun phrases are complementary to verb phrases)" (Ibid.). As one might expect by now, a referential concept is to be exercised jointly with a predicable concept in a speech or mental act to inform the act with a referential nature, just as the predicable concept informs that act with a predicable nature (Ibid.).

At this stage, it is tempting to examine a concrete example. Let us use Cocchiarella's example, the case in which the quantifier phrase contains what he calls a sortal concept: 'All swans are white'. The assertion 'All swans are white' has cognitive structure that corresponds to its grammatical analysis as '[All swans]NP [are white]VP'. Where S is a sortal constant for 'swan', and W is a monadic predicate constant for 'white', the logical form of the assertion is represented by Cocchiarella's theory as $(\forall xS)[\lambda xW(x)](x)$. But exactly what are being accomplished by the noun phrase or the quantifier phrase?

Cocchiarella is meticulous and audacious enough to indicate what amounts to the identifying description (if not the definition) of referential concept:

> Referential concepts, on our analysis, are what quantifier phrases stand for when the latter are properly represented as containing a type of expression that we call names, both proper and common, and both simple and complex. By a common name we mean here a common count noun, including those (such as 'cat', 'dog', 'tree', 'animal', etc.) whose use in thought and communication is associated with certain identity criteria, as well as those (such as 'object', 'thing', or 'individual') that are not associated with any specific identity criteria. (Ibid., 174)

This description is striking for several aspects which induce the following protests. First of all, how could referential concepts be what quantifier phrases stand for? Secondly, how could they be what quantifier phrases stand for when the latter are properly represented as containing names? Thirdly and finally, how could that be the case when proper names are contained?

In the first two questions, we are asking how a quantifier phrase could stand for anything, let alone referential concepts. In other words, even if we accept everything Cocchiarella says about rewriting ordinary quantifier phrases $(\forall x)$ and $(\exists x)$ as abbreviated forms of $[\forall xObject]$ and $[\exists xObject]$ in his theory of logical form, we shall need more to make sense of Cocchiarella's idea that referential concepts are what quantifier phrases stand for.

I think the situation is somewhat similar to the situation in which we were puzzled by Frge's claims about so-called second-level functions, i.e., "functions whose argument are and must be functions" (Frege 1952, p. 38). In a case like "$\neg (\forall x)\neg f(x)$", we are supposed to have a function whose argument is indicated by 'f'. I do not believe that it is easy to understand how the second level function, which is unsaturated, could be saturated with another function, which is also unsaturated. Frege himself conceded that by saying that "such a function is obviously a fundamentally different one from those we have dealt with so far". Clearly there is a fundamental difference between functions whose arguments are and must be functions and functions whose arguments are objects and cannot be anything else. However, Frege was parsimonious in giving further informative characterizations as to how they are fundamentally different. For example, we are not informed whether in both cases we have the same process of saturation. Since the saturation of a first-level function with an object must be fundamentally different from that of a second-level function with another function, we have every right to ask Frege to elaborate the difference. After all, what is the point of differentiating the two cases, in case we are devoid of such an elaboration?

As was reported above, Cocchiarella claims that his notion of unsaturatedness is similar but different from that of Frege's. Even though it is far from complete, Cocchiarella reveals some hints regarding how he views the differences between their notions of unsaturatedness. Interestingly, such a revelation seems to be found in Cocchiarella's answer to our third question, to which we now turn.

3.2 Singular Reference of Proper Names

According to Cocchiarella, proper names also have identity criteria, and in that respect they are not unlike sortal common names. And their criteria "are determined for the most part by a common name sortal that corresponds to the proper name" (Ibid., 177). So, Cocchiarella takes "proper names to be a type of sortal common names that necessarily satisfy a condition of uniqueness":

> That is, a sortal name S is a proper name only if S can be used to refer to at most one thing, or, in symbols, only if $(\forall xS)(\forall yS)(x = y)$ is a conceptual (necessary) truth. (If modal operators are added to our theory, we assume that proper names are "rigid", so that $(\forall xS)\Box(\forall yS)(x = y)$ is also a necessary truth. (Ibid., 178)

Cocchiarella agrees with Geach regarding the point that both proper and common names are different from predicate expressions in that they can be used outside the context of a sentence in simple acts of naming. It is interesting to note, however, Cocchiarella does not count such use of names as the use of a name to refer:

> Naming is not the same as referring, it should be emphasized, because the latter is an act that does not occur outside the (implicit if not explicit) context of a sentence used in a speech act, i.e., independently of an act of predicating (Cocchiarella 1999; reprinted in Cocchiarella 2007, pp. 169–214)

Cocchiarella's emphasis on the distinction between naming and referring reminds of the medieval distinction between *significatio* and *suppositio* (see Park 1997b, pp. 207–230). Be that as it may, by focusing on the logical form of the referential use of a proper name, Cocchiarella is now able to contrast efficiently his own theory of singular reference with other theories of the 20th century:

> ...the representation of a proper name in logical syntax is to be none other than as an individual constant – just as Frege and most of the contemporary views, including the so-called "new" theory of direct reference, would have it.
>
> On our account, proper names are no different from common names in the way they are used to stand for referential concepts, and, in particular, in the way they inform a speech or mental act with a referential nature. (Cocchiarella 1998, 177)

4   Frege and Cocchiarella

Given Cocchiarella's theory of proper names, i.e., singular reference, we may better understand the difference between his notion of unsaturatedness and that of Frege's. Unlike Frege's case, where a saturated object was needed to fill the gap of an unsaturated concept (or, in general, function), Cocchiarella requires mutual saturation of a referential concept and a predicable concept, both of which are unsaturated.

As was reported above, Frege himself noticed the difference between the saturation of the first level function whose argument is an object and saturation of the second level function whose argument is a function. Why did he not understand the difference in terms of different types of saturation, i.e., the former as a saturation from an saturated object and an unsaturated function, and the latter as a saturation from an unsaturated function's filling the gap of an unsaturated second-level function. If he had done so, he could not have arrived at what Cocchiarella calls the mutual saturation of a referential concept and a predicable concept. As one can detect from the context of discussion and the apologetic tone in Frege's remarks, we may think that Frege felt uneasiness when he noticed the difference. Why? It is probably because he was worried about the danger of the shipwreck of his system as a whole due to the troublesome difference. And, probably for that reason, he introduced the infamous "value-ranges" (see Frege 1952, p. 32). As far as I know, as I pointed out elsewhere (see Park 1990, 1992), Cocchiarella is the first person who claimed that Frege equated value-ranges with concept-correlates, which are the denotata of nominalized predicates. The point is that value-ranges, concept-correlates, and the denotata of nominalized predicates are all objects in Frege's ontology. That means, Frege's notion of saturation needs objects without any exception even by paying the high price of introducing the dubious category of value-ranges.

Cocchiarella seems to be alert about the importance of emphasizing the difference between his notion of mutual saturation and Frgee's notion of saturation. For he writes:

It is just this sort of mutual saturation of complementary cognitive structures that constitutes the nexus of predication in conceptualism. It is also what accounts for the unity of a speech or mental act, i.e., of an assertion or judgment, a problem that Ockham, who anticipated F. H. Bradley's infinite regress argument, was unable to resolve. Ockham, for example, assumed that a judgment that every man is an animal was literally made up of a universal quantifier, the concept *man*, the mental copula is, and the concept *animal*. But then what unifies these mental terms into a single unified mental act? A fifth mental term that "tied" these terms together would need a sixth to "tie" it with the others, which in turn would need a seventh, and so on ad infinitum. That is not how a judgment or assertion is understood in conceptual realism, where concepts, as unsaturated cognitive structures, are not objects, and therefore cannot be actual constituents of a mental act (event). (Cocchiarella 1999, p. 19; Cocchiarella 2007, p. 179)

Furthermore, in a footnote, he even claims that on his account of predication, "there cannot be even a first step toward Bradley's infinite regress".

Elsewhere, Cocchiarella also contrasts the difference between his notion of saturation and that of Frege's with somewhat different emphasis as follows:

We adopt the terminology of unsaturatedness here from Frege, but differ from in taking concepts to be cognitive structures and not independently real functions from objects to truth values. The objectivity of a concept (as an intelligible universal) consists not in its independence to mind and cognition, according to conceptualism, but in its intersubjectivity as a cognitive capacity, i.e., in the fact that different people can exercise the same concept at the same time (or at different times), as well that the same person can exercise the same concept at different times. (Cocchiarella 1997, 180)

All these contrasts of Frege and Cocchiarella are of utmost importance from ontological as well as cognitive perspective. In his discussion of the logical equivalence of quantifier phrases in his theory of logical form and ordinary quantifiers, his account of complex common names with his new operator for attaching a relative clause, his distinction between reference with and without existential presupposition, and his discussion of active versus deactivated referential concepts, all of which are suppressed in this article, Cocchiarella is anxious to emphasize the superiority of his theory in that it represents the cognitive structure of our speech and mental acts. For example, he writes:

It is this distinction between logical forms that represent the cognitive structure of our speech and mental acts and logical forms that represent the truth conditions of those acts that is important and fundamental in conceptual realism; and it is a distinction that is not to be found in standard theories of reference, including the so-called "new" theory of direct reference. (Cocchiarella 1998, 179)

Certainly, Cocchiarella has a very good reason for emphasizing that his theory is a conceptual theory, i.e., "a theory that attempts to explain the use of referential concepts in speech and mental acts, which is what some philosophers of language call a pragmatic theory, as opposed. E.g., to a purely abstract semantical theory" (Ibid., 170). For, again as he himself points out, "this is where reference has its basis and primary role, and that any other notion of reference will be derivative upon this" (Ibid.).

However, I believe, Cocchiarella should have contrasted his notion of saturation and that of Frege's with slightly different emphasis, as far as his primary concerns centers around the problems of reference. In fact, Cocchiarella not only draws our attention to the paradigm of reducing general reference to the singular reference of logically proper names that dominated the 20th century but also discussed all relevant aspects of Frege's philosophy, in which the notion of saturation looms large.[8] Everything points to a working hypothesis that, by failing to anticipate Cocchiarella's notion of the mutual saturation of a referential concept and a predicable concept, Frege led us all to the wrong direction of reducing general reference to the singular reference of logically proper names. Probably, Cocchiarella's own theory of reference may lead us to the opposite direction of reducing singular reference to the general reference of variable binding term operators. The only step left for Cocchiarella is to name Frege as the originator of the paradigm. What is going on?

No doubt, Cocchiarella's study of (onto)logic cannot get off the ground without Frege's pioneering achievements. It must have been a tormenting struggle for Cocchiarella to nail down his intellectual father's small error as the original sin for the paradigm of reducing general reference to the singular reference of logically proper names. Who knows, however, Cocchiarella is securing a more proper and more dignified place for Frege in the history of philosophy adjacent to Aristotle and the medieval logicians?

5 Conclusion

In this chapter, I presented a sketch of the most rudimentary part of Cocchiarella's theory of general as well as singular reference. Though it is not easy to understand even how his theory works, it must be worthwhile to reflect upon its significance against the background of the entire history of the theories of reference. By discussing the difference between Cocchiarella's notion of saturation and that of Frege's, I believe, I did a bit of what should be done in that direction. Insofar as

---

[8]In addition to the writings of Cocchiarella already mentioned, Cocchiarella (1988) seems most important in this connection. See also, Cocchiarella (1992).

recent discussions as to the origin of the new direct theory of reference (see Humphreys and Fetzer (eds.) 1998) is within the purview of analytic philosophers, we may safely predict the imminent surge of interest among them about the origin of what Cocchiarella calls the paradigm of reducing general reference to the singular reference of logically proper names.

References

Cocchiarella, N. B. (1986). *Logical investigations of predication theory and the problem of universals*. Naples: Bibliopolis Press.

Cocchiarella, N. B. (1987). *Logical studies in early analytic philosophy*. Columbus, Ohio: Ohio State University Press.

Cocchiarella, N. B. (1988). Predication versus membership in the distinction between logic as language and logic as calculus. *Synthese, 77*, 37–72.

Cocchiarella, N. B. (1992). Conceptual realism versus quine on classes and higher-order logic. *Synthese, 90*, 379–436.

Cocchiarella, N. B. (1995). Knowledge representation in conceptual realism. *International Journal of Human-Computer Studies, 43*, 697–721.

Cocchiarella, N. B. (1997). Conceptual realism as a theory of logical form. *Revue Internationale de Philosophie, 2*, 180.

Cocchiarella, N. B. (1998). Reference in conceptual realism. *Synthese, 114*, 169–202.

Cocchiarella, N. B. (1999, 2001). A logical reconstruction of medieval terminist logic in conceptual realism. *Logical Analysis and History of Philosophy, 4*, 35–72.

Cocchiarella, N. B. (2007). *Formal ontology and conceptual realism*. Dordrecht: Springer.

Frege, G. (1952). *Translations from the philosophical writings of Gottlob Frege*. In P. Geach & M. Black (Eds.). Oxford: Basil Blackwell.

Humphreys, P., & Fetzer, J. (Eds.). (1998). *The New theory of reference: Kripke, Marcus, and its origins*. Dordrecht: Kluwer.

Park, W. (1990). Scotus, Frege, and Bergmann. *The Modern Schoolman, 67*, 259–273.

Park, W. (1992). Frege's distinction between 'falling under' and 'subordination' (in Korean). *Yonsei Philosophy, 4*, 53–80; reprinted in Park (1997a), pp. 49–74.

Park, W. (1996). The theory of causation in medieval Aristotelianism. In Korean Society for Analytic Philosophy. (Ed.), *Causality and theories of causation*. Seoul: Philosophy and Reality Press; reprinted in Park (1997b), pp. 283–308.

Park, W. (1997a). *In search of the lost science*. Seoul: Damron Sa. (in Korean).

Park, W. (1997b). *The seduction of medieval philosophy*. Seoul: Philosophy and Reality Press. (in Korean).

Park, W. (2001). On Cocchiarella's retroactive theory of reference. In O. Majer (Ed.), *Logica yearbook 2000* (pp. 79–89). Prague: Czech Academy of Science.

Part II The Hilbert School

Chapter 5
Zermelo and the Axiomatic Method

Abstract This chapter intends to examine the widespread assumption, which has been uncritically accepted, that Zermelo simply adopted Hilbert's axiomatic method in his axiomatization of set theory. What is essential in that shared axiomatic method? And, exactly when was it established? By philosophical reflection on these questions, we are to uncover how Zermelo's thought and Hilbert's thought on the axiomatic method were developed interacting each other. As a consequence, we will note the possibility that Zermelo, in his early as well as late thought, had views about the axiomatic method entirely different from that of Hilbert. Such a result must have far-reaching implications to the history of set theory and the axiomatic method, thereby to the philosophy of mathematics in general.

Keywords Axiomatic method · Zermelo · Hilbert · Deepening the foundations
Implicit definition · The concept of set

1 Introduction

As is well-known, Zermelo built the first axiomatic system of set theory in 1908. ZFC is undisputably the privileged standard axiomatic set theory of our time. If so, in view of the special role and status of set theory as the foundation of entire mathematics, Zermelo's philosophical views must be of our focal interest. Surprisingly, however, it is rare to find detailed philosophical study of Zermelo.

There are some understandable reasons for this state of affairs. First, there is a widespread interpretation, according to which what motivated Zermelo to axiomatize set theory was simply to avoid Russell's paradox. Secondly, according to this interpretation, Zermelo simply adopted Hilbert's axiomatic method. Thirdly, as a mathematician (just as usual for mathematicians), Zermelo rarely presented his philosophical views. Finally, some commentators have claimed that, even if we can find in the later Zermelo some unique views about the ontological status of

This chapter was originally published as Park (2008) in Korean.

mathematical objects, which are different from that of Hilbert, they are clearly different from the problems of methodology.

However, there are many serious problems in this stereotypical ways of understanding. Above all, Russell's paradox and the alleged resulting crisis in mathematics have been too much emphasized. As a consequence, it has been rare to understand some other potentially more important motivations for Zermelo's axiomatizing set theory. It is, of course, true that Hilbert's *Foundations of Geometry* (1899) was the birth certificate for modern history of the axiomatic method. Further, Hilbert stood at the center of modern discussions of the axiomatic method by his proof theory and metamathematics. Nevertheless, Zermelo's axiomatization of set theory may not be merely an instance of applying and realizing the Hilbertian ideal of the axiomatic method. We should bypass the possibility that Zermelo's axiomatic system of set theory might have played a central role in the development of study of the axiomatic method. Even if it is difficult to find apparently philosophical discussions in Zermelo, we cannot set aside the possibility of extracting his philosophical ideas from the content of his axiomatic set theory itself. The controversy between him and his critics among mathematicians could be the enough source for understanding his philosophical position. Further, the implicit assumption that it is possible to distinguish sharply between the ontological problem from the methodological problem itself is dubious. At least, we should not forget the possibility that the difference in the former might bring up the difference in the latter.

If we reflect upon the reason why Zermelo's views on the axiomatic method failed to attract much attention, one salient point that looms large is that indeed the four reasons discussed above are intricately interrelated, and ultimately due to the lack of balanced understanding of the similarities and the differences between Hilbert and Zermelo. From this point of view, to question what specific motivations, methodological views, and philosophical position Zermelo had could not only be satisfying our historical curiosity but also a crucial clue for understanding and pursuing the most advanced on-going researches in set theory and philosophy of mathematics. Even if we simply maintain the standard views, according to which the crisis in mathematics was caused by Russell's paradox, the importance of the critical reexamination of Zermelo's achievement that appeased the criticism cannot be too much emphasized.

Fortunately, there have been recent attempts to reevaluate Zermelo's thought based on posthumous writings, correspondences, and unpublished lecture notes. It is a challenge for historians that due to chronic illness Zermelo hardly published articles in 1910s and 1920s. What was happening to Zermelo's views on the axiomathic method in those years? More interesting and even mysterious is the fact that Hilbert neither focused on the foundations of mathematics in the years between 1908 and 1917 nor commented on Zermelo's system of axiomatic set theory. Of course, recent scholarship has improved the situation enormously. At least, we can witness that some of the obvious misunderstandings and prejudiced interpretations have been questioned. However, there are certain limitations in these attempts, for they tend to confine the difference between Hilbert and Zermelo in ontological issues by assuming that there is no serious disagreement between them as far as the axiomatic method is concerned. What is the essence of the axiomatic method assumed to be shared by them? Exactly when was it established? It would be unfortunate, if we implicitly assume that there was no remarkable change in Hilbert's thought about the axiomatic method since he published Foundations of Geometry and until he initiated proof theory. For, by avoiding these questions, we might give up opportunities to get an in-depth understanding of one of the most exciting cases of revolutionary change in the history of mathematics. This article, through philosophical reflections, aims at an explication of how Zermelo and Hilbert developed their views on the axiomatic method with active interactions.

Here is my strategy. In Sect. 2, I will report some clues gathered from recent researches that discuss the differences between Hilbert and Zermelo found in the later Zermelo's papers written around 1930. Of course, I shall discuss these differences selectively depending on whether they could be useful for identifying the essence of the axiomatic method. In Sects. 3 and 4, I shall inquire whether the differences in their views about the axiomatic method are not confined to the later Zermelo's thought but traceable to much earlier periods. For this purpose, in Sect. 3, I shall first highlight the fact that Hilbert's thought went through enormous changes in both his attitude toward set theory and his position about the axiomatic method in the period between 1899 and 1917. First, I shall summarize the usual interpretation of Hilbert's vie about implicit definition as an essential element that shouldn't be vulnerable to such a change. Then, I shall cast a doubt as to whether we can sharply distinguish between the traditional axiomatic approach since Euclid and Hilbert's axiomatic method, and suggest that Frege/Hilbert controversy has not been ended. Also, I will discuss Hilbert's view that the application of the axiomatic method to diverse fields of mathematics and natural sciences is at the same time deepening the foundations of a given field. Finally, what changes Zermelo-Russell paradox brought up to Hilbert's thought will be discussed briefly. In Sect. 4, based on Zermelo's twin papers published in 1908, Zermelo's position on the axiomatic method will be fathomed against the background of Hilbert's views discussed in Sect. 3. I shall discuss first Moore's interpretation, according to which what motivated Zermelo for his axiomatization of set theory was not Zermelo-Russell paradox but his desire to fortifying his own well-ordering theorem. Then, I shall argue that the conflict between Zermelo's reductionistic tendency and Hilbert's model theory was not merely stemming from their different philosophical views but necessarily accompanying some methodological differences. On that ground, I shall estimate Hilbert's and Zermelo's positions on the status of set theory as the foundation of mathematics. Also, by discussing the necessary preconditions for set theory as a particular mathematical discipline to be axiomatized, I shall interpret Zermelo's work as deepening the foundations of Cantor's set theory. Naturally, we will be able to argue that Zermelo's position lies somewhere between the previous Euclidean approach and Hilbert's axiomatic method. The concluding part of Sect. 4 will be devoted to the issue of determining whether there is Zermelo's unique thought distinct from that of Hilbert about implicit definition, for it is the ultimate task of this chapter. In Sect. 5, finally, I will hint at some future agenda, merely touched upon here.

2   The Discrepancy Between the Later Zermelo's Thought and Hilbert Program

Interestingly and remarkably, Hilbert publicly expressed his views on Zermelo's 1908 achievement, i.e., the first axiomatic set theory, in his famous Zürich lecture in 1917, only after a long period of silence. There can several explanatory hypotheses for the possible cause of that silence. For example, one may appeal to the fact that at that time there was a heated debates on the axiom of choice, or another fact that Hilbert was preoccupied with the problem of axiomatizing physics, or still another fact that, even though quite short, Hilbert was attracted to Russell's logicism. However, there is no doubt that until 1917 the relationship between Hilbert and Zermelo, whether it be personal or academic, was a very friendly one.

However, the situation seems to have changed drastically in the period between the later 1920s and the early 1930s, when Zermelo was once more contributed to the foundations of mathematics actively. Recently, Peckhaus in an article with such an interesting title as "Pro and Contra Hilbert: Zermelo's Set Theories" highlights this point (see Peckhaus 2005). According to him, Zermelo in this period opposed to almost all positions in the contemporary foundations of mathematics. In particular, he did not follow Hilbert's move toward metamathematics. While Hilbert, in championing metamathematics, gave up the traditional neutrality in epistemology and ontology, thereby suggesting a constructive foundations of mathematics approaching to Brouwer's intuitionism, Zermelo rejected any kind of finitistic approach to mathematics as the expression of Skolemism in set theory. As an alternative, according to Peckhaus, Zermelo sustained his idealistic approach by trying to justify the infinite hierarchy of sets in terms of what he called "the logic of the infinite" (ibid., 3.2).

Peckhaus' report about the differences between Zermelo and Hilbert, contrary to the title of his paper promises, is too brief. It merely points out some modes of the conflicts between their opinions without any attempt to uncover and explain what caused such a conflict. Furthermore, we cannot gather from his report any information as to what such differences imply for understanding the axiomatic method. In that regard, Hallett's report is much more useful for my present purpose, for he makes it explicit not only that Zermelo in the later period deviated radically from Hilbert's standpoint but also that that fact reveals nothing but the faults of the axiomatic method.

As an evidence for that fact that Zermelo deviated from Hilbert's position toward the axiomatic systems, Hallett quotes the following passage from Zermelo (1930):

> Our axiom system is non-categorical, which in this case is not a disadvantage but rather an advantage, for on this very fact rests the enormous importance and unlimited applicability of set theory. .... It is customary to confine attention to the smallest infinite domain, the "Cantorian", and I see little advantage in doing this. Rather, set theory as a science must be developed in the fullest generality, and then the comparative investigation of individual models can be undertaken as a particular problem. (Zermelo 1930, p. 45; Hallett 1995b, p. 63)

Also, Hallett suggests an interpretation that Zermelo's discussion revealed the problems of Hilbert's axiomatic method based on the following quotation:

> Scientific reactionaries and anti-mathematicians have so eagerly and lovingly appealed to the "ultrafinite antinomies" in their struggle against set theory. But these are only apparent "contradictions" and depend solely on confusing *set theory* itself, which is not categorically determined by its axioms, with individual *models* representing it. What appears as an "ultrafinite non- or super-set" in one model is, in the succeeding model, a perfectly good, valid set with both a cardinal number and an ordinal type, and is itself a foundation stone for the construction of a new domain. To the unbounded series of Cantor ordinals there corresponds a similarly unbounded double series of essentially different set theoretic models, in each of which the whole classical theory is expressed. The two polar opposite tendencies of the thinking spirit, the idea of creative advance and that of collections and *completion* [*Abschluss*], ideas which also lie behind the Kantian "antinomies", find their symbolic representation and their symbolic reconciliation in the transfinite number series based on the concept of well-ordering. This series reaches no true completion in its unrestricted advance, but possesses only relative stopping points, just those "boundary numbers" which separate the higher model types from the lower. Thus, the set theoretic "antinomies", when correctly understood, do not lead to a cramping and mutilation of mathematical science, but rather to an, as yet, unsurveyable unfolding and enriching of that science. (Zermelo 1930, p. 47; Hallett 1995b, pp. 63–64)

In order to appreciate the full implication of Hallett's views in these quoted passages, we need to invoke the entire history of the axiomatic method from Zermelo (1908a) to 1930, which is simply beyond the scope of this paper. In view of my modest aim to extract some clues for understanding Zermelo's early thought from his later work, what we need to note is that, in confronting Skolem's criticism that axiomatic set theory is not absolute, Zermelo seems to suggest implicitly the pitfalls of Hilbert's axiomatic method. In other words, Zermelo seems to blame Hilbert's for such problems in order to safeguard his axiomatic set theory and set theories in general. However, such a move seems to conflict with the usual understanding that Zermelo tried to axiomatize set theory in the spirit of Hilbert's axiomatic method.

3 The Development of Hilbert's Thought on Axiomatic Method Between 1900 and 1917

As was pointed out above, it is widely assumed that in his axiomatization of set theory Zermelo adopted Hilbert's axiomatic method. Such an assumption seems to be fully supported by Zermelo's autobiographical acknowledgement of his indebtedness to Hilbert, Hilbert's praise of Zermelo's axiomatic set theory as the exemplary application of the axiomatic method, and all circumstantial evidences including their personal relationship. However, the fact that the later Zermelo's thought betrays remarkable differences from Hilbert program provokes our curiosity as to whether Zermelo's early thought is in agreement with Hilbert's thought in the same period.

3.1   *Implicit Definition in Hilbert*

Where are we to find the definite character of Hilbert's axiomatic method clearly distinct from the axiomatic method in the broad sense, which can be trace back at least to the axiom systems of Euclidean geometry? Usual answer to questions like this seems to be suggested by appealing to the view that the meaning of the primitive terms is suggested by the implicit definition. For example, F. A. Muller claims that Hilbert counts getting "something between explicit definition and having no definition, namely having an *implicit definition*" (Muller 2004, 429). He quotes from Hilbert's letter to Frege in this regard:

> In my opinion, a concept [primitive notion, FAM] can be fixed logically only by its relations to other concepts. These relations, formulated in certain statements, I call *axioms*, thus arriving at the view that axioms (perhaps together with propositions assigning names to concepts) are the *definitions* of the concepts. (Hilbert in Frege 1980, p. 51; Muller 2004, 429. Muller's emphasis)

Also, Muller pints out that von Neumann explicitly adopted the same view in his axiomatization of Cantorian set theory:

> By 'set' we understand here (in the sense of the axiomatic method) something of which one does not know anything more *and does not want to know anything more* than what follows from the axioms. (Neumann 1925, 36, our translation; Muller 2004, 429)

As Muller does, we arrive at this stage the idea that "a notion is *implicitly definable* iff we can axiomatize it" (ibid.). Then, he explains that here in implicit definition there is "a *definiendum* without *definiens*", and instead "a variety of rules of how to use the new concept in combination with other concepts".[1]

3.2   *Euclidean Axiom System and Hilbertian Axiomatic Method: Unending Frege/Hilbert Controversy*

If we understand Hilbert's axiomatic method as the towering achievement in the history of axiomatic method, it is an almost universal custom to treat Hilbert's *Foundations of Geometry* (1899) as the watershed, thereby assuming that the essentials of Hilbert's axiomatic method was presented in a completed form there. Then, the concept of implicit definition as the core of Hilbert's axiomatic method, as was sketched in the previous subsection in accordance with the usual understanding, must have been sustained continuously from its birth in 1899. For, if the

---

[1]Further, Muller claims that, at least in mathematics, Wittgenstein's 'social' conception of meaning is harmonious with Hilbert's 'rational' conception of implicit definability: "because accepted axioms of some branch of mathematics, together with the logical deduction-rules, govern the rigorous *use* of the primitive notions as they are actually *used* by the community of mathematicians in this branch" (Muller 2004, 429).

concept of implicit definition were to be abandoned, it would be meaningless to distinguish between the traditional Euclidean axiomatic approach and the axiomatic method as the original invention of Hilbert. Also, even if the concept of implicit definition itself has not been abandoned, if there were some drastic conceptual changes, at least some very careful attention to different historical periods would be required in order for us to talk about "Hilbert's axiomatic method".

However, did Hilbert stick to his views on the axiomatic method in *Foundations of Geometry* (1899) until 1917 without any slight change? Are Hilbert's views, to which Zermelo opposed in his later thought, exclusively due to the changes brought up in the process of responding in collaboration with Bernays to the criticisms of intuitionists and Weyl, and digesting the achievements of Skolem? Is the axiomatic method embedded in Hilbert (1899), which is claimed to be followed by Zermelo in his first axiomatic set theory of 1908, the same one as the axiomatic method meant by Hilbert in 1920, when he praised Zermelo (1908b) as the example of the axiomatic method? And, is that allegedly same method the method of defining the ultimately fundamental concepts in a given area implicitly by axioms?

Majer seems to share my suspicions, for he claims that

something *must be wrong* with the 'received view': "Hilbert separated himself from Euclid's axiomatic approach and aimed at a completely new conception of axiomatics". (Majer 2006, p. 158)

According to Majer, when Hilbert first gave a lecture on geometry in 1891 "he didn't know in which *direction* he should steer his investigations" (ibid.). It was his reading of Pasch's book that made Hilbert to be clear about his own "principal goal",

i.e., "to revive Euclid's axiomatic point of view, which had been badly neglected during the eighteenth and nineteenth century, and bring geometry in this manner to a new logical perfection". (Majer 2006, p. 157)

So, Majer emphasizes that Hilbert's axiomatic method did not exist at the outset but "emerged bit by bit *during* his investigations" (ibid.). If so, I think, Majer is absolutely right in his assessment of our current situation: "the real significance of Hilbert's efforts to revive Euclid's axiomatic approach was neither correctly apprehended nor fully understood" (ibid.).

After more than a century's effort, do we now understand clearly the axiomatic method as the method of modern mathematics? Though we have been brainwashed that, as far as the axiomatic method is concerned, there is a huge difference between before and after Hilbert, are we merely fallen into an obscurantism, without understanding either Hilbert's axiomatic method or the traditional Euclidean axiomatic approach, or without any effort to understand any? If we reflect on these matters encouraged by Majer's questions, we may realize how long we have postponed to settle Frege/Hilbert controversy (see Resnik 1980; Hallett 1994; Shapiro 1997). For, as is well known, Frege's complaints, doubts, and criticisms against Hilbert were exactly from the point of view of the traditional Euclidean axiomatic approach to the aspects of Hilbert's new axiomatic method.

For example, Frege complained that Hilbert did not provide us with definitions (e.g., definition of "between"), and axioms are instead carrying the burden that belongs to definitions. Then, Frege claimed that a definition should specify the meaning of word whose meaning is not yet given, that a definition should use other words whose meanings are already known, that, unlike definitions, axioms and theorems should not include words or signs whose sense and meaning or what it contributes to the expression of certain thought, and that axioms should express truth (see "Frege to Hilbert (December 27, 1899)", Frege 1980, pp. 34–38). In connection with this letter, Shapiro summarizes the issues as follows:

> According to Frege, axioms should express *truths* and definitions should give the *meanings* and *fix the denotations* of certain terms. With an implicit definition, neither job is accomplished. (Shapiro 1997, p. 161)

Hallett, by quoting Bernays, also emphasizes that, unlike the standard views on axiomatic systems before Hilbert, axioms for Hilbert are neither truths nor judgments that can be either true or false. Furthermore, even the entire axiom system does not express truth (Hallett 1995a, 137; Bernays 1922, 95–96). Hallett thinks that such a difference in viewing whether axioms have truth values originates from the difference in viewing the status of axioms. If we assume that axioms are true, since they would be truths about the relations among the primitive terms that appear in axioms, universal logic that could determine those primitive terms would be pursued. On the other hand, "if the assumption about the axioms as truth is dropped, "there is no necessity to say anything about the primitives prior to the development of the theory" (Hallett 1995a, 141). Also, we can gather from Hallett very useful discussions on the determinacy of the reference. For example, according to Hallett, "reference independence" Hilbert ascribed to non-logical constant amounts to rejecting Frege's theory, in which sense determines reference. Further, Hilbert thought that reference independence should also be applied to the problem of truth for the axioms (Hallett 1994, 163).

It is beyond the scope of this chapter to look into the details of Frege/Hilbert controversy. The significance of Frege/Hilbert controversy for my present purpose can be found, however, for the following two reasons. First, In order to establish Hilbertian axiomatic method clearly distinct from the traditional axiomatic approach since Euclid, persuasive replies should be given to the objections to Hilbert in Freg/Hilbert controversy.[2] Secondly, contrary to the impression from his replies to Frege, if there were some changes, if any, in Hilbert's views on the axiomatic method due to the controversy, we should pay careful attention to them.[3]

---

[2]Rowe (2000, p. 76): "While Hilbert failed to respond to Frege's detailed critique, his silence should not be taken as a sign that he failed to comprehend the point of Frege's criticisms".

[3]Peckhaus (1994, p. 99): "Above all, it is remarkable that in this passage Hilbert upholds the traditional view that axioms are propositions, since back in 1899, in a famous controversy with Frege over the "Grundlagen der Geometrie", Hilbert already advanced the "modern" notion that the word "axiom" does not denote a proposition but rather a propositional function".

In order to justify the first reason, perhaps it would be enough to point out that the issues formulated by Shapiro are still effective. For example, the idea that the axioms should be true is not only intuitively natural but also adopted by those, like Muller, who defend implicit definition, without special discussion (see Muller 2004, 439). In order to justify the second reason, it would be enough to remember the following facts Resnik pointed out in his discussion of Frege/Hilbert controversy:

> In Hilbert and Bernays's *Grundlagen der Mathematik*, Volume I (1934), there is a description of the axiomatic method which (except for avoiding higher-order logics) agrees with Frege's account of Hilbert's reduction of geometry to pure logic. (Resnik 1980, p. 113)[4]

What is implied by all this is rather obvious. Hilbert's position apparently expressed in his correspondence with Frege may not faithfully represent his mind. It could be merely betraying Hilbert's emotion in competitive rivaly. It seems highly likely that Hilbert may have experienced significant inner conflict and changing positions. In other words, Hilbert's true position seems to sustain more elements of the traditional Euclidean axiomatic approach than we ascribe to Hilbert as the founder of the new Hilbertian axiomatic method.

3.3   The Application of the Axiomatic Method and Deepening the Foundations

Another important problem we cannot ignore in discussing Hilbert's axiomatic method is how to secure in-depth understanding of Hilbert's epistemological and methodological perspectives on human knowledge as a whole and the place and the role of mathematics within it. For this purpose, we need to understand Hilbert's thought about the applications of the axiomatic method in diverse fields of pure mathematics and natural sciences. But that would require to discuss several problems, including the problem of selection of axioms in particular disciplines, the problem of deepening the foundations, the problem of understanding the differences between pure mathematics and natural sciences in applying the axiomatic method, and the problem of axiomatizing logic itself. If we just remember the interaction between Einstein and Hilbert, we may easily imagine how actively Hilbert experimented on the axiomatization of diverse scientific fields in the years between 1899 and 1917.[5] Also, Hilbert in this period was deeply immersed in philosophical problems embedded in the axiomatic method through his friendly relationship with

---

[4]In fact, Resnik even evaluates Frege as follows: "In any case, Frege was the first to give written evidence of an adequate grasp of the use of implicit definitions" (Ibid.).

[5]See Rowe (2000, p. 82): "Yet while retreating on the foundational front, Hilbert widened his interest in axiomatic dramatically during the twelve-year period from 1905 to 1917. This period marked at once the pinnacle of achievement for Göttingen mathematics as well as for "Hilbert's school".

many people around him such as Nelson and Zermelo. There is no doubt that the debates with Brouwer and Weyl or Zermelo-Russell paradox must have deeply stimulated Hilbert and left lasting influences to him. In a word, in the middle of a turmoil, where revolutionary and monumental achievements were flowing out and competing with each other at various levels and directions, Hilbert was a centering figure in many ways in all these events.

Hilbert (1918) could be counted as a personal final report of such a scholarly experiment. Here, Hilbert emphasizes the central role of mathematics among the sciences. There we find the following suggestive claims:

> The procedure of the axiomatic method, as it is expressed here, amounts to a deepening of the foundations of the individual domains of knowledge—a deepening that is necessary for every edifice that one wishes to expand and to build higher while preserving its stability. (Hilbert 1918, p. 1109)

This sentence, which seems to claim that deepening the foundations of the individual science necessarily demands its expansion to its neighbouring fields of knowledge, provides us with an extremely interesting issue. In fact, the reexamination of Hilbert's views on the foundations of physics is quite in vogue in recent researches on Hilbert. We can only welcome this trend in view of the well-known fact that the problem of axiomatizing physics is the sixth of the twenty-three problems presented by Hilbert in 1900 (see Hilbert 1900b). For example, Majer's discussion contains the potentiality of serving as the springboard for further fruitful research:

> ...the application of the axiomatic method in physics is only legitimate insofar as one takes it as an expansion (or transfer) of the axiomatic point of view in geometry. This does not imply that they have to be exactly the same; there may be certain differences due to the differences between geometry and physics. On the other hand, Hilbert was deeply convinced (from the very beginning of his occupation with geometry) that geometry is a natural science, indeed the most fundamental of all natural sciences, and this is the deeper reason why the axiomatic point of view should be transferable to physics. (Majer 2006, p. 164. Majer's emphasis)

Considering Hilbert (1918), I would now like to confirm the continuity of the Euclidean axiomatic approach and Hilbert's axiomatic method by highlighting selectively what is implied by Majer's observations. At the same time, I will probe the question as to what differences the one and the same axiomatic method would make depending on the peculiarity of the individual science that is to be axiomatized. Hopefully, we may thereby fathom the desiderata for axiomatization of any given field of knowledge. Immediately before the sentences quoted above, Hilbert explained by exploiting ample examples how the axiomatic method is applied and how the deepening of the foundations results in a given field of knowledge. Let us follow Hilbert's lead to capture the major stages in a schematic fashion.

> (1) Collected facts in a given field (of knowledge) can be ordered by a certain conceptual framework.

(2) In establishing the conceptual framework for the field (of knowledge), there are some underlying salient propositions. (The propositions are themselves good enough for building the entire framework according to logical principles.)

(3) The fundamental propositions can be considered as axioms for the individual field (of knowledge). (The development of the field depends only on the additional logical establishment of the conceptual framework.)

- -

(4) The problem of establishing the foundations of the field (of knowledge) has been solved.

(5) However, the solution is only temporary. A need arises to ground the propositions themselves that are fundamental and axiomatic in the field (of knowledge) (Based on Hilbert 1918, pp. 1107–1109; see Park 2008, pp. 19–20).

It is not so difficult to follow Hilbert's reasoning here, but the problem is whether it is also easy to understand the following emergent thought. How to ground those fundamental propositions themselves? Hilbert's answer is as follows:

So one acquired 'proofs' of the linearity of the equation of the plane and the orthogonality of the transformation expressing a movement, of the laws of arithmetical calculation, of the parallelogram of forces, of the Lagrangian equations of motion, of Kirchhoff's law regarding emission and absorption, of the law of entropy, and of the proposition concerning the existence of roots of an equation. (Hilbert 1918, p. 1109, [7])

However, Hilbert is quick to point out that those proofs are not in themselves proofs.

But critical examination of these 'proofs' shows that they are not in themselves proofs, but basically only make it possible to trace things back to certain deeper propositions, which in turn are now to be regarded as new axioms instead of the propositions to be proved. The actual so-called axioms of geometry, arithmetic, statics, mechanics, radiation theory, or thermodynamics arose in this way. These axioms form a layer of axioms which lies deeper than the axiom-layer given by the recently-mentioned fundamental theorems of the individual field of knowledge. (Hilbert 1918, p. 1109, [8])

If we add Hilbert's explanation like this to the scheme discussed above, then we can have the following stages:

(6) We try to trace those fundamental propositions to certain deeper propositions (by proofs that are not proofs in a rigorous sense).

(7) Those deeper propositions are now regarded as new axioms.

(8) Those new axioms form a layer of axioms which lies deeper than the axiomlayer given by the recently-mentioned fundamental theorems of the individual field of knowledge (Based on Hilbert 1918, p. 1109; see Park 2008, p. 21).

Now, in light of the comprehensive perspective provided by Majer, let us reflect on Hilbert's explanation just quoted and its schematic summary. The difficulties in understanding Hilbert's deep thought that the application of the axiomatic method is nothing other than the deepening of foundations are due to the fact that there are overlaps of several possibilities of confusions. For convenience, let us suppose that it is possible to distinguish between applying the axiomatic method to particular areas of knowledge and expanding one and the same method to the neighboring areas. There are two problems that might cause confusions in understanding Hilbert's thought on the axiomatic method. The first, which has been detected by Lowe, Majer, and Stöltzner, is <1> the problem of confusion, which might creep in when we fail to distinguish between launching an axiomatic system and the application of the axiomatic method. The second is <2> the problem of confusion that appears when we apply the axiomatic method to a particular area and deepening the foundations, already the expansion to neighboring disciplines is involved (even if we assumed that they are distinguishable at least for argument's sake). In understanding Hilbert's thought about expanding the axiomatic method to neighboring areas, confusion can be brought up due to the fact that <3> "expanding the axiomatic method to neighboring area" may be interpreted either as meaning the application of the same method or meaning that, in an individual case where we apply the axiomatic method to a certain field, neighboring disciplines may be involved.

It seems that Problem <1> has recently been relatively well covered by several scholars, thereby proved the usefulness of the distinction involved. In order to get off the ground, axiomatization must presuppose some already accumulated knowledge in a given area. For, if not, it would be almost meaningless to question the dependence and independence among the propositions in the given area of knowledge. If so, the stage at which ample knowledge has been accumulated, and an axiom system has suggested by the Euclidian axiomatic approach, must be a stage, at which we do not yet have reduced the axioms to other new axioms. In other words, it is a stage, at which axioms themselves have not been justified, deepening the foundations has not been realized, and Hilbertian axiomatic method has not yet been applied. Thus, both the reason why Hilbert (1918) discussed the problems of independence (or dependence) and consistency, and the reason why several scholars have emphasized the importance of distinguishing between the presentation of an axiom system and the axiomatic method, can be found in the importance of deepening the foundations by applying Hilbertian axiomatic method.

One historical fact, notable when we focus on the poin that the application of the axiomatic method is the deepening the foundations, is that scholars in Hilbert's school, such as Nelson, heatedly discussed the problem of regressive method for a long period of time (see Peckhaus 2002). It is not difficult to find some traces of such discussions even in Hilbert (1918). For, the method by which we discover deeper propositions that will acquire the status of axioms by proofs that are rigorously speaking non-proofs, could be this regressive method. The method of analysis and synthesis can be traced back to Pappus and the ancient Greek geometrical analysis. This method was led to the method of resolution and composition in the Middle Ages. It was also treated importantly in the Modern period by distinguished philosophers such as Descartes, Newton, and Kant. Though intriguing, it is obviously beyond the scope of this article. Let it suffice to point out that there is plenty room for raising substantial issues, even only in the context of Hilbert's axiomatic method, such as "Can this regressive method be continued infinitely?", or "Is this regressive method bound to be continued infinitely?"

In view of their importance, problem <2> and problem <3> surprisingly have been rarely discussed. Notable, and extremely meaningful exception seems to be Stöltzner (2002), where he analyzed the cases of deepening the foundations in diverse individual sciences suggested in Hilbert (1918) by categorizing them into eight different types (see Stöltzner 2002, pp. 256–259).

3.4 The Influence of Zermelo-Russell Paradox

Interestingly, Peckhaus claims that the paradoxes of set theory brought about important philosophical changes in Hilbert's early axiomatic program. As evidence, he cites the fact that, unlike the early Hilbert who did not understand the logical relevance of the paradoxes, set theory now became in Hilbert the focus of his project of axiomatizing mathematics (Peckhaus 1994, p. 92).[6] Peckhaus also points out that Hilbert did not take the axiomatic approach to set theory (Peckhaus 1994, p. 96).[7] In Hilbert, set theory was a program that works together with his foundations of arithmetic, and he claimed that within the framework of his axiomatic all troublesome concepts of set theory are "either not expressible or their existence can be proved". Further, Peckhaus claims that around 1900 Hilbert treated set theory not as an independent discipline but "rather as an alternative methodological approach to arithmetic". According to Peckhaus,

> [i]n accordance with his "pragmatic" viewpoint, the appearance of contradictions was nothing to be alarmed about so long as the stock of accepted mathematical knowledge could be preserved by other means. (Peckhaus 1994, p. 96)

Then, suddenly by 1905, Hilbert completely changed his attitude toward set theory. In a newly developed field of mathematics, i.e., set theory, some unresolved problems were found. They even evoked contradiction. So, "[t]heir foundations, therefore, urgently needed to be strengthened and their methods thoroughly investigated" (Peckhaus 1994, p. 97).

Further, Peckhaus emphasizes the point that the paradoxes influenced

> the germ of his programme, the consistency proof for the axioms of arithmetic, because it was impossible to obtain the desired "direct" proof by means of an inconsistent logic. (Ibid.)

As a consequence, unlike his previous view that treated the "axiomatization of arithmetic as a purely mathematical problem, independent of logical and

---

[6]Moore also points out the same point as follows: "Although in 1900 Hilbert was aware of various paradoxes of set theory, he did not show any concern that set theory, or logic, was threatened". Moore (2002, 47).

[7]Moore also gives us an interesting report in the same vein. According to the lecture Notes in Summer, 1905, even though Hilbert suggested the use of the axiomatic method, he did not suggest the axiomatization of set theory (Moore 2002, 49).

set-theoretic investigations", Hilbert came to require a "partly simultaneous development of the laws of logic and arithmetic" (Ibid., p. 95).

After pointing out that, even though the first two problems in Hilbert's 23 open problems of mathematics concerned set theory, only in 1908 they came to be discussed in the context of Zermelo's axiomatic set theory, Rowe makes an interesting comment:

> Even then, since Zermelo's axioms failed to specify any logical rules for set formation, his system had no real use when it came to answering Hilbert's key questions: can a direct proof be found establishing the consistency of arithmetic? (Rowe 2000, 82)[8]

Above in Sect. 3.3., based on Hilbert (1918), we considered Hilbert's views on applying the axiomatic method to diverse scientific disciples. Now, against that background, we turn to examine Hilbert's views presented in the same lecture on the application of the axiomatic method to a special disciple, i.e., set theory. In the context, where he discussed the problem of the consistency of axioms in connection with deepening the foundations, Hilbert pointed out first that evidently the problem of consistency has utmost importance. And, he found the reason in that the existence of contradictions endangers the content of the entire theory. Interestingly, he noted that, while the contradictions appearing in physical theories are eliminated always by changing the selection of axioms, in the case of the contradiction appearing in the field of purely theoretical disciplines, the situation is entirely different:

> Set theory contains the classic example of such an occurrence, namely, in the *paradox of the set of all sets*, which goes back to Cantor. This paradox is so serious that distinguished mathematicians, for example, Kronecker and Poincaré, felt compelled by it to deny that set theory—one of the most fruitful and powerful branches of knowledge anywhere in mathematics—has any justification for existing. (Hilbert 1918, 1112, [31])

Then, Hilbert made an extremely interesting claim for our present purpose:

> But in this precarious state of affairs as well, the axiomatic method came to the rescue. By setting up appropriate axioms which in a precise way restricted both the arbitrariness of the definitions of sets and the admissibility of statements about their elements, Zermelo succeeded in developing set theory in such a way that the contradictions disappear, but the scope and applicability of set theory remain the same. (Hilbert 1918, 1112, [32])

If we read this paragraph as isolated, readers may well get the impression that, after a long period of silence without any public announcement after Zermelo's 1908b axiomatic set theory, Hilbert finally expressed his agreement and support to it. Also, such an impression seems shared by the experts. For example, Ebbinghaus writes in the first biography of Zermelo as follows:

> The axiomatization paper is the keystone of Zermelo's set-theoretic work during his first period of foundational research. As argued above, it is written in the spirit of Hilbert's

---

[8]Rowe (2000, p. 82). Rowe conjectures that for that very reason, even though Hilbert supported Zermelo strongly, he did not go out and play any active role in the controversy of Zermelo's axiom of choice. As we already pointed out that Hilbert's silence is a kind of mystery, We should welcome Rowe's conjecture, whether it is right or wrong.

foundational programme. Hilbert himself belonged to its admirers. In his 1920 course he says ([Hil20], 21–22):

> The person who in recent years has newly founded [set] theory and who has done so, in my view, in the most precise way which is at the same time appropriate to the spirit of the theory, that person is Zermelo. (Hilbert 1920, 21–22; Ebbinghaus 2007, p. 79)

And, he characterizes the axiomatization as

> the most brilliant example of a perfected elaboration of the axiomatic method [through which] the old problem of reducing all of mathematics to logic gains a strong stimulus. (in the same lecture 33; Ebbinghaus 2007, p. 79)

However, I think that such an interpretation is wrong. Above all, it is problematic that that the source, on which Ebbinghaus depends, is not a published document but merely a lecture note. To say the least, we have to count Hilbert (1918), which was a lecture later published in a journal, more prestigious than lecture notes. Now, if we consider the context in Hilbert (1918), in other words, to read the texts before and after the quoted paragraph together with it, we may get an entirely different impression. Let us read first the following passage in the quotation:

> In all previous cases it was a matter of contradictions that had emerged in the course of the development of a theory and that needed to be eliminated by a reformation of the axiom system. But if we wish to restore the reputation of mathematics as the exemplar of the most rigorous science it is not enough merely to avoid the existing contradictions. The chief requirement of the theory of axioms must go farther, namely, to show that within every field of knowledge contradictions based on the underlying axiom-system are *absolutely impossible*. (Hilbert 1918, 1112, [33])

This paragraph clearly claims that it is not enough to have what Zermelo achieved in Zermelo (1908b). Let us take a look at the prior paragraph, where the problem of paradoxes was dealt with:

> As one can see from what has already been said, the contradictions that arise in physical theories are always eliminated by changing the selection of the axioms; the difficulty is to make the selection so that all the observed physical laws are logical consequences of the chosen axioms. (Hilbert 1918, 1112, [30])

Wasn't Hilbert here suggesting that the way Zermelo avoided contradictions is similar to the way we eliminate contradictions in physics by changing the selection of axioms in physics? If so, wasn't Hilbert implicitly suggesting that Zermelo should be criticized for his insufficient understanding of the application of the axiomatic method in that he ignored the fact that, in case set theory is taken to be the foundation of mathematics, we should take a way of eliminating contradictions entirely different from the way of eliminating contradictions in physics?

After discussing the so-called problem of relative consistency in the subsequent paragraphs, Hilbert claims as follows:

> In only two cases is this method of reduction to another special domain of knowledge clearly not available, namely, when it is a matter of the axioms for the *integers* themselves,

and when it is a matter of the foundations of *set theory*; for here there is no other discipline besides logic which it would be possible to invoke.

But since the examination of consistency is a task that cannot be avoided, it appears necessary to axiomatize logic itself and to prove that number theory and set theory are only parts of logic. (Hilbert 1918, 1113, [38, 39])

Hilbert's claim reconfirms his assessment of limitations of the achievement of Zermelo's 1908b article, and it is consonant with the relatively critical tone of Hilbert(1918) over Zermelo (1908b). What is interesting is that now Hilbert turned to the project of axiomatizing logic in order to solve the paradoxes. Moore describes on this matter as follows:

This statement shows that, contrary to the usual interpretation, Hilbert was then a logicist— in the sense that he believed mathematics can be reduced to logic. (Moore 2002, 54)

However, again as is pointed out by Moore, it seems that Hilbert's logicist years didn't last long. For, in his lecture in the Summer of 1920,

[h]e concluded that "today the goal of reducing set theory to logic (and thereby also reducing the customary methods of analysis to logic) has not been achieved, and perhaps cannot be achieved in general". (Hilbert 1920, 33; Moore 2002, 57)

What is remarkable is that in this lecture, contradistinction to his critical tone in Hilbert (1918), Hilbert heartily praised Zermelo (Ibid.). In a word, it seems that Hilbert was fluctuating between Zermelo's axiomatic set theory and Russell's logicism. In other words, since he realized freshly the meaning of Zermelo-Russell paradox, Hilbert ascribed to the problem of solving or resolving it an utmost importance.

4   Zermelo's Position on the Axiomatic Method Around 1908

4.1   The Motivations for the Axiomatization

As we noted above, the motivation for Zermelo to axiomatize set theory has been found usually in the set-theoretic paradoxes (see Sect. 3.4). According to Moore, however, "[s]uch a view contains a morsel of truth but ignores essential differences between Zermelo and his contemporaries" (Moore 1982, p. 157). Here, Moore is comparing Russell and Hausdorff with Zermelo. According to Moore, while Russell tried to solve the paradoxes by theory of types as a theory of underlying symbolic logic, Haudorff tried to find the foundations of the ordered sets within mathematics. And, Zermelo's position lies between them:

Unlike Russell, in 1908 Zermelo was not preoccupied with the paradoxes (popular mis-conception to the contrary), but with the reception of his 1904 proof. Yet, like Russell, he realized that Cantor's theory of the transfinite needed some thoughtful pruning and, specifically, that Cantor's definition of a set—as any totality of well-distinguished objects of our thought—was untenable. Both Hausdorff and Zermelo considered set theory to be a part of mathematics, rather than of philosophy or logic.* However, in contrast to Hausdorff, Zermelo regarded an axiomatization as essential to the sound future development of set theory (* Zermelo 1908b). (Moore 1982, p. 158)

Moore further notes that in both Zermelo (1908b), that presented a new proof of the well-ordering theorem, and Zermelo (1908a), that presented the first axiomatic set theory, paradoxes were treated as only an apparent threat (Moore 1982, p. 158). It is Moore's interpretation that what was threatened was rather Zermelo's proof of well-ordering theorem, and Zermelo presented a "two-fold answer":

First, he replied to his critics at length and gave a new demonstration, which still depended as heavily as the first on the Axiom of Choice. Second, he created a rigorous system of axioms for set theory and embedded his proof within it. He knew that in order to preserve the entire proof, such a system had to include his Axiom of Choice. Thus his axiomatization was primarily motivated by a desire to secure his demonstration of the Well-Ordering Theorem and, in particular, to save his Axiom of Choice. (Moore 1982, p. 159)[9]

There must be a grain of truth in both the majority view that finds the motivation for Zermelo's axiomatization of set theory in Russell's paradox and the minority view that finds it in Zermelo's desire to justify his own proof of well-ordering theorem. At the same time, they are not incompatible either. If so, we should set aside the desire to identify one of the two as the only or crucially important motivation for a while. It is necessary first to give effort to find the motivation for Zermelo's axiomatization of set theory at a deeper level, so that we can assign appropriate place to both Hilbert and Zermelo without failing to differentiate between them. In other words, the project of axiomatizing set theory was common to both Hilbert and Zermelo. The resolution of Russell's paradox and the proof of well-ordering theorem were merely the necessary preconditions for realizing the project. We should start from these to move on to the inquiry asking each of Hilbert and Zermelo had what basic motivation for their unique axiomatization project of set theory.

We have already relatively ample clues. For example, as Hilbert already pointed out in his letter to Frege, Russell's paradox was discovered several years ago by Zermelo (Zermelo 1908b, 118–119; Moore 1982, p. 159). Since Hilbert came to consider it seriously only after realizing that it is a logical paradox not confined to set theory (Peckhaus 1994, pp. 94–98), the significance of Russell's paradox in Hilbert should be found more in the problem of boundary between logic and mathematics rather than in the problem of the axiomatization of set theory. Also, since the problem of the proof of well-ordering theorem is a part of the problem of Cantor's continuum hypothesis, the first among 23 Hilbert problems (Hilbert 1900a, pp. 1103–1104), Zermelo's desire to justify his own proof can be easily located in the process of specifying the idea of axiomatizing set theory. However,

---

[9]Of course, one might treat this as a minority opinion. However, we should note that Moore's interpretation has been accepted by significant number of scholars including Hallett, Peckhaus, and Kanamori.

what does it mean to axiomatize set theory? Even if both Hilbert and Zermelo tried to axiomatize set theory, couldn't they mean entirely different thing by "*the axiomatization of set theory*"? Above we distinguished between building an axiomatic system and the reflection of the axiomatic method itself. Also, we contrasted building an axiomatic system and the application of axiomatic method in the sense of deepening the foundations. Was Hilbert, in Hilbert (1918), evaluating Zermelo's axiomatic set theory as merely building an axiomatic system without any reflection on the axiomatic method itself or any application of the axiomatic method in the sense of deepening the foundations?

4.2   Zermelo's Reductionism

To understand the rivalry between the majority view and the minority view about the basic motivation for the axiomatization of set theory as the rivalry between Hilbert's and Zermelo's views about the meaning of axiomatizing set theory seems to be a very promising working hypothesis. Now, we need to estimate whether we may understand these differences as the differences in views about the status of set theory within the kingdom of science or within mathematics, in other words, the difference between set theory as a field of mathematics and set theory as foundation of mathematics. Let us find our point of departure from R. G. Taylor, who is a rare case that pointed out the differences between Hilbert and Zermelo around 1908. For, as Taylor points out, there seems to be a serious irreconcilable conflict between Zermelo's reductionistic tendency and Hilbert's model-theoretic ways of understanding (Taylor 1993, 543–545).[10]

Taylor understands reductionism as consisting of the following three theses:

(R1) All mathematical objects are sets.

(R2) All mathematical concepts are definable in terms of membership.

(R3) All mathematical truths are set-theoretic truths.

Also, he posits a much more moderate view, which rejects the three theses above, but accepts the following three similar theses:

(R'1) All mathematical objects may be understood as sets.

(R'2) All mathematical concepts may be understood in terms of membership.

(R'3) All mathematical truths may be understood as set-theoretic truths.

---

[10]Kanamori (2004, 499f) discusses Zermelo's reductionism as follows: "Zermelo pioneered the reduction of mathematical concepts and arguments to set-theoretic concepts and arguments from axioms, based on sets doing the work of mathematical objects". Zermelo (1909) wrote at the beginning: "... for me, every theorem stated about finite numbers is nothing other than a theorem about *finite sets*". See also Hallett (1984, p. 244f).

Taylor claims that, even though there is no explicit statement amounting to (R1)–(R3) in Zermelo's early papers, Zermelo's aim is clearly to develop a foundational program, and this program is reductionistic in character. And, he thinks that Zermelo did not explicitly state (R1)–(R3), because he counted Cantor and Dedekind as already having proved reductionism (at least) for finite numbers (Taylor 1993, 540).

Interestingly, Taylor's arguments for thinking that Zermelo simply followed Hilbert's methodology is very weak, for his only evidences are merely that Zermelo was concerned for his failure to prove the consistency of his axioms in Zermelo (1908a), and that he also mentioned the problem of completeness (Taylor 1993, 542–543).

On the other hand, Taylor's arguments in favor of the fact that Zermelo had very different philosophical position from Hilbert are relatively strong and to the point. For, he declares that what he understands to be Hilbert's philosophical position as an early prototype of model-theoretic viewpoint, and presents persuasively the reasons why Zermelo's position cannot agree with it. Taylor claims that, according to Hilbert, axiom systems have models, and this led him to new important ideas regarding mathematical existence and truth. For, if various models are given, it would be meaningless to talk about the mathematical objects that exist independently of the domain of the model. Similarly, it would be meaningless to speak of true propositions that are true independently of those models. On the other hand, Taylor points out that, in Zermelo, the variety of the models are extremely limited, for, according to Zermelo, domain can vary only for the urelement. Further, Taylor emphasizes that Zermelo wanted to sustain the tradition view, according to which axioms of mathematics are self-evident truths. While Zermelo never avoided the problem of truth, and defended the view that mathematics is an a priori science based on self-evident truths, Such views cannot be adopted from Hilbert's point of view (Taylor 1993, 543).

However, Taylor understands these differences merely as differences in philosophical positions, and claims that there was no differences in their methodologies (ibid.). I think such a difference in philosophical positions, due to its far-reaching implications for the method of selecting axioms, necessarily entail the differences in methodological views.[11] From this point of view, what is urgent is to argue for the point that the methodological difference will be revealed in the axioms Zermelo actually adopted and their roles. Also, it would be meaningful to have an overview of how Hilbert understood and digested such conflicts.

Some people might try to block the possibility itself that the differences between the philosophical views of Hilbert and Zermelo imply their methodological differences, by separating sharply the problem of the selection of axioms in a given field and the problem of the reflection about the axiomatic method itself. They

---

[11]In fact, everywhere in Taylor's article, those that can be counted as the methodological differences are pointed out as the difference between Hilbert and Zermelo. For example, see Taylor (1993, 537). This point will be further discussed in what follows.

might argue: Even if the reductionistic tendency in Zermelo, through its ample implications to the problem of the existence of mathematical objects and the problem of mathematical truth, brought up a sort of methodological differences between Hilbert and Zermelo, that would hardly be a hindrance for the basic agreement of Hilbert's and Zermelo's view about the axiomatic method itself.

However, I believe that such an objection would distort the actual mathematical and philosophical practice of Hilbert, and posit a fictional position very narrowly understood. As we saw above in Sect. 3, Hilbert showed remarkable flexibility in applying the axiomatic method to the various areas of pure mathematics and natural sciences. This reminds us of the fact that Hilbert was not confined to either the problem of selecting axioms and the problem of the reflection on the axiomatic method itself, so that the problem of epistemology or methodology was also Hilbert's focal interest.

4.3   The Problem of Selecting Axioms and the Axiom of Choice

The point of departure for detecting Zermelo's epistemological or methodological views can be found in his lecture notes for the first mathematical course in Germany in the Summer of 1908.[12] There we find his discussion of "Are propositions of arithmetic analytic or synthetic?" While there was almost universal agreement about propositions in geometry that they originate from intuitions to have the character of synthetic judgments, in the case of arithmetic, there was an on-going debates between those who opted for analytic character such as Frege, Russell, and Peano, and those who opted for synthetic character such as Poincaré. According to Ebbinghaus' report, instead of siding with either, Zermelo assumed that both analytic and synthetic judgments can appear in arithmetic, and aimed to sort out analytic elements. And, the method used there seems to be the regressive method mentioned above (Ebbinghaus 2007, pp. 97–98).

Another source we can rely on is the Introduction of Zermelo (1908a). Though brief, it provides us with a few significant hints. First of all, the first sentence of this paper announced that set theory is the logical foundation of arithmetic and analysis:

> Set theory is that branch of mathematics whose task is to investigate mathematically the fundamental notions "number", "order", and "function", taking them in their pristine, simple form, and to develop thereby the logical foundations of all of arithmetic and analysis; thus it constitutes an indispensable component of the science of mathematics. (Zermelo 1908a, p. 200)

---

[12]Zermelo's own manuscript and a note taken by Kurt Grelling are preserved (Ebbinghaus 2007, p. 97).

As Moore points out efficiently, there were many mathematicians who did not treat set theory as the appropriate foundation of the entire mathematics even in the period around 1917:

> Some of them were suspicious of set theory, such as many French mathematicians, including some who had used parts of it in their research (e.g., René baire and Henri Lebesque). Some rejected set theory as a foundation and rejected mathematical logic at the same time, as did L. E. J. Brouwer and Henri Poincaré. Some rejected set theory as a foundation because the proper foundation of mathematics was mathematical logic; here the most striking representative was Bertrand Russell. And many, perhaps most mathematicians at the time showed no concern with foundational questions at all. (Moore 2002, 52)

Also, as was already discussed, in the case of Hilbert, partial and simultaneous development of logic and arithmetic was proposed in Hilbert (1905), and, in the two exceptional cases of arithmetic and set theory, where relative consistency proofs are impossible, the axiomatization of logic itself was suggested in Hilbert (1918). Thus, it was truly an exceptional, ambitious, controversial, and prophetic event that announced set theory as the foundation of mathematics in 1908.

Secondly, Zermelo report that the very existence of set theory is threatened by particular contradictions or "antinomies" derivable from principles governing our thought necessarily, and satisfactory solution had not been found.

> In particular, in view of the "Russell antinomy" (1903, oo. 101-107 and 366-368) of the set of all sets that do not contain themselves as elements, it no longer seems admissible today to assign to an arbitrary logically definable notion a set, or a class, as its extension. (Zermelo 1908a, p. 200)

Then, he noted that Cantor's definition of set requires a slight constraint. Though this is not particularly informative, it enables us to estimate Zermelo's perception of the background situation.

Thirdly, Zermelo describes the only way out in such a situation for him as follows:

> Under these circumstances there is at this point nothing left for us to do but to proceed in the opposite direction and, starting from set theory as it is historically given, to seek out the principles required for establishing the foundations of this mathematical discipline. (Ibid.)

This sentence seems to have enormous potential, for it could mean entirely different thing depending on, opposition to what was meant in proceeding "on the opposite direction", and what Zermelo intended to refer to by "set theory as it is historically given". Hallett interprets "set theory as it is historically given" from the point of view that understands Zermelo's axiom of separation as a principle of restricting the size of set to make precise the definition of set suggested by Cantor in 1882. According to him, even though Dedekind, Frege, and Russell used universal domain as a consequence of their preference to logical simplicity,

> [B]ut when this use of a universal domain had clearly broken down (one could call it a logical over-simplification) it was not surprising that those interested in producing an axiomatization of mathematical set theory (as opposed to set-theoretic logic or logicistic mathematics) should want to return to and clarify the original limited operation with sets. (Hallett 1984, p. 242)

And Hallett equates here the concept of the original limited operation with set with Scott (1974)'s "our original intuition of set is based on the idea of having collections of already fixed objects" (Scott 1974, p. 207; Hallett 1984, p. 242). However, this interpretation seems to be slightly problematic in that it without special discussion interprets "set theory as it is historically given" as "set theory as was originally presented", i.e., "set theory as was originally presented by Cantor". This interpretation has a merit that allows us to understand easily as "to proceed in the opposite direction" as "to proceed in the opposite direction of the direction took by Dedekind, Frege, and Russell". But, shouldn't the set theories of Dedekind, Frege, and Russell have the rights to be considered as equally as that of Cantor's as historically presented? I consider that such an interpretation would be the right way of understanding "as it is historically given".

In other words, by interpreting "set theory as it is historically given" as literally as "set theory as it is historically presented", so that it could be the entirety of mathematical truths (or theorems) accumulated centering around Cantor, Dedekind, Frege, and Russell, and we should interpret it in that way. Of course, in that case, the price to pay is that we need to find out the meaning of "to proceed in the opposite direction". However, it seems to me that answer to this problem is already at hand, contrary to what one might think. At the beginning of Sect. 4.3., we took Zermelo's lecture notes to have an overview of his epistemological and methodological views. As we saw, there he, instead of taking side with the rival views in the controversy about whether proposition of arithmetic are analytic or synthetic, assumed that in arithmetic both analytic and synthetic judgments can appear, and aimed at sorting out analytic elements only. Then, we immediately surmised that his method of sorting out the analytic elements was nothing but the so-called "regressive method". Now we can think that that regressive method is the method of moving from the given mathematical propositions in the opposite direction to find out the fundamental propositions that are required to be necessarily true in order for the given mathematical propositions to be true. So, I claim that we may understand Zermelo's stream of thought as follows: "We started from the principles necessarily governing our thought and Cantor's definition of set, and deduced some particular contradictions or "antinomies". We know that some constraints should be given to the concept of set itself, but we have failed to find the alternative definition of set. So, what we can do at this stage is nothing else than to move starting from the truths of set theory as it was historically given in the opposite direction to find out the principles required to establish the foundations of this area of mathematics".

Fourthly, Zermelo, in his article, made it explicit that what he intended is to show

> how the entire theory created by Cantor and Dedekind can be reduced to a few definitions and seven principles, or axioms, which appear to be mutually independent. (Zermelo 1908b, p. 200)

It is an unmistakably important clue that he mentioned Dedekind in this context. However, the most important point seems to be that Zermelo was using explicitly the term "reduction" here.

Fifthly, Zermelo enumerated what he would not discussed in the article:

The further, more philosophical, question about the origin of these principles and the extent to which they are valid will not be discussed here. (ibid.)

"[T]he origin of these principles" seems to be somewhat ambiguous. This could mean the more ultimate principles, on which these principles are based. Or, by asking how we discovered the principles, it could mean the origin of the ideas. On the other hand, philosophical problems about "the extent to which they are valid" may be safely understood as the boundary problem between logic and mathematics.

Next, we need to confirm whether we can gather clues from what Zermelo had to say about the axioms of his own axiomatic set theory in his twin articles in 1908. As is well known, the axiomatic system of Zermelo (1908a) consists of 7 axioms: (1) extensionality, (2) foundations, (3) separation, (4) power set, (5) union, (6) choice, (7) infinity. Though all these deserve extensive discussion, historians usually focus on separation, choice, and infinity, with reasonable ground.[13] First of all, regarding the axiom of separation, Zermelo claimed that it (by granting freedom to a certain extent in defining a new set) provides us with a substitute of the Cantorian definition of set. And, he found its difference from Cantor's definition in that there are two significant restrictions. The first restriction that "sets may never be *independently defined* by means of this axiom but must always be separated as subsets from sets already given" eliminates contradictory notions such as "the set of all sets" or "the set of all ordinals". On the other hand, the second restriction that the defining criterion must always definite resolves Richard antinomy. Finally, Zermelo wrote that, even though it would be enough to have the axioms enumerated so far in order to derive all essential axioms of set theory, he adopted the axiom of infinity due to Dedekind in order to guarantee the existence of infinite set.

Insofar as the axiom of choice is concerned, Zermelo (1908a) does not contain any special suggestive discussion of the axiomatic method. On the other hand, Zermelo (1908b), when he categorized the criticisms of his proof of well-ordering theorem, provide us with much information invaluable for our purpose in the context of responding to the criticisms raised against his use of the axiom of choice. In particular, after having pointed out that even Peano's *Formulaire* (1897), which was an attempt to reduce the whole mathematics to "syllogism" (in Aristotelian-scholastic sense), relies on significant number of principles that are impossible to prove, immediately protested as follows:

First, how does Peano arrive at his fundamental principles and how does he justify their inclusion in the *Formulaire*, since, after all, he cannot prove them either? Evidently by analyzing the modes of inference that in the course of history have come to be recognized as valid and by pointing out that the principles are intuitively evident and necessary for science—considerations that can all be urged equally well in favor of the disputed principle. That this axiom, even though it was never formulated in textbook style, has frequently been used, and successfully at that, in the most diverse fields of mathematics, especially in set theory, by Dedekind, Cantor, F. Berstein, Schoenflies, J. König, and others is an

---

[13]For example, see Moore (1982).

indisputable fact, which is only corroborated by the opposition that, at one time or another, some logical purists directed against it. Such an extensive use of a principle can be explained only by its *self-evidence*, which, of course, must not be confused with its provability. No matter if the self-evidence is to a certain degree subjective—it is surely a necessary source of mathematical principles, even if it is not a tool of mathematical proofs, and Peano's assertion (1906a, p. 147) that it has nothing to do with mathematics fails to do justice to manifest facts. But the question that can be objectively decided, whether the principle is *necessary for science*, I should now like to submit to judgment by presenting a number of elementary and fundamental theorems and problems that, in my opinion, could not be dealt with at all without the principle of choice. (Zermelo 1908b, pp. 187–188)

It is obvious that we can find the essence of Zermelo's thought about the axiomatic method in this quoted paragraph. Above all, it is remarkable that, in the situation where the axiom of choice is under attack, he was expressing his confidence on its self-evident character. For, the fact that Zermelo appealed to the self-evidence of axioms can be a clear evidence that he sustained the traditional Euclidean views rather than following Hilbert's axiomatic method regarding such an important issue. Of course, he did not claim that mathematical truth or proposition is self-evident. However, when he said that "mathematical principles (or axioms) are self-evident", nevertheless, if he had not been treating those principles (or axioms) as true proposition (or proposition that can be true or false), what could he mean by that? The point that mathematical principle (or axiom) is necessary for science is also significant in other respects. For, we can be curious about its relevance to indispensable argument, which is one of the most important issues in recent philosophy of mathematics. It is not clear whether Zermelo considered intuitive self-evidence and necessity for science as a necessary condition, sufficient condition, or necessary and sufficient condition for axioms. What is rather obvious is that Zermelo was presenting an ad hominem argument against the opponents in the argumentative context. It is not my concern whether it commits an informal fallacy, or allowable in such an argumentative context. The only reason why I am invoking this point is to draw attention to the fact that both Zermelo and his opponents were debating over the status of the axiom of choice by assuming the traditional Euclidean axiomatic approach. Perhaps, if there were any who would object to the assumption at that time, he or she could have been only Hilbert. If so, in other words, Zermelo, while he was axiomatizing set theory in the spirit of Hilbert's axiomatic method, claimed about the status of axioms what Hilbert would certainly dislike or protest.

4.4 Could Zermelo's Axiomatic System of Set Theory Be the Implicit Definition of the Concept of Set?

As we saw above in Muller's case, and just as usual, if we reason by assuming that Zermelo, under the influence of Hilbert adopt Hilbert's axiomatic method to axiomatize set theory, it is natural to understand that Zermelo defined set implicitly by his axiom system. Though rare, however, it is not the case that there was no case in which such an understanding met criticism. For example, Taylor claims that interpreting Zermelo's intention as model-theoretic in the Hilbertian sense, and thereby understanding his 7 axioms as defining a particular class of Cantorian structures, is attractive but fallacious (Taylor 1993, 544). Also, long time ago, Poincasré pointed out that if Zermelo considered his axioms as implicit definition of the term "set", there was a need to prove the consistency of them. However, according to Poincaré, Zermelo never did it, and couldn't do it, for, if he were to do so, he would have to rely on pre-established truths. However, it is not possible, for Zermelo wanted that his axioms would be entirely autonomous as foundations of mathematics (Poincaré 1909, 473; Requoted from Moore 1982, p. 161).

Now, based on such antecedents, I shall try to construct an argument that might confirm the impossibility that Zermelo's system of axiomatic sst theory cannot be an implicit definition of set concept.

Argument 1 (Poincaré argument)

Consistency proof is a necessary condition for a given axiomatic system to be an implicit definition of certain fundamental term. But Zermelo failed to prove the consistency of Z. Therefore, the axioms of Z cannot be the implicit definition of set concept.

I think that this argument is conclusive. And, this point is also found in Hilbert (1918), where he pointed out the limitations of Zermelo's axiomatic set theory. Further, Zermelo himself in Zermelo (1908a) expressed disappointment for his failure to prove consistency.

Then, even if the axioms of Z cannot be the implicit definition, why do we treat them as if they are implicit definitions? Probably, the only possible answer would be that we believe that they were intended to be implicit definitions.

Only as for the case in which there is no slight doubt for our belief that certain axioms are intended to implicitly define certain primitive concepts simultaneously, even if the consistency of the axiom system had not been proven, let us count them as implicit definitions. In such a case, could the axioms of Z be implicit definition of a set? I think that we have sufficient enough reasons for doubt it. As we saw above, among the axioms of Z, most of the role of characterizing the concept of set is taken care of by the axiom of separation. Indeed, Zermelo himself mentioned that point. Thus, we might construct an argument as follows:

Argument 2

If axioms of a certain axiom system are counted as an implicit definition, there shouldn't be any slight doubt as to our belief that the axioms of Z were intended to define simultaneously and implicitly the primitive terms. But there are a few points to believe that the axioms of Z were intended to be an implicit definition. Therefore, the axioms of Z cannot be counted as an implicit definition of the concept of set.

Even if they cannot be an implicit definition, and even if they cannot be counted as an implicit definition, is there still a possibility that the axioms of Z are still an implicit definition? I am extremely suspicious about that point.

5   Concluding Remarks

Understanding Zermelo's conception of sets must be presupposed in order to secure an in-depth understanding of his views about the axiomatic method, which was revealed through his axiomatization of set theory. As a consequence, some readers might think that I should have discussed the following important problems: (1) Hallett's view that Zermelo basically adopted Russell's idea of avoiding the paradox by limiting the size of sets, and Taylor's criticism of that view, (2) the view, which was spread widely through Gödel, that modern axiomatic set theory implicitly presupposes Zermelo's iterative conception of set, and (3) Skolem/ Zermelo controversy caused by Skolem paradox, and the problem of first order vs. second order logic. However, I believe, there are good excuses for avoiding these problems here. AS for (1), Insofar as Zermelo did not count the problem of paradox as so important, it would be just a subsidiary issue to find how he got the idea of avoiding it. As far as (2) and (3) are concerned, since it is simply impossible for them to influence on Zermelo's thought around 1908 or his thought between 1899 and 1917, which is the focus of this chapter, we can safely reserve them for a future study on Zermelo's thought in later periods.

Hilbert's axiomatic method is one of the issues Zermelo never touched upon in Zermelo (1908a). Didn't he perhaps to show by the paper itself that, at least for set theory, Hilbert's axiomatic method is not applicable? In this chapter, I considered the uncritically and widely accepted assumption that Zermelo adopted Hilbert's axiomatic method in axiomatizing set theory. As a consequence, it was confirmed that there is a possibility that not only in his later thought but also in his earlier thought Zermelo might have views on set theory and the axiomatic method entirely different from that of Hilbert's. This result is bound to have far-reaching implications to the history of set theory, the history of axiomatic method, and the philosophy of mathematics.

References

Bernays, P. (1922). Ueber Hilberts Gedanken zur Grundlegung der Arithmetik. *Jahresbericht der Deutschen Mathematiker-Vereinigung, 31*, 10–19 (English translation in Mancosu (1998a), pp. 215–222).

Ebbinghaus, H.-D. (in cooperation with V. Peckhaus). (2007). *Ernst Zermelo: An approach to his life and work*. Berlin: Springer.

Ewald, W. B. (Ed.). (1996). *From Kant to Hilbert. A source book in the foundations of mathematics* (Vol. 2). Oxford: Oxford University Press.

Frege, G. (1980). *Philosophical and mathematical correspondence* (G. Gabriel et al., Eds.). Chicago: The University of Chicago Press.

Hallett, M. (1984). *Cantorian set theory and limitation of size*. Oxford: Clarendon Press.

Hallett, M. (1994). Hilbert's axiomatic method and the laws of thought. *George, 1994*, 158–200.

Hallett, M. (1995a). Hilbert and logic, in Marion and Cohen (1995), pp. 135–187.

Hallett, M. (1995b). Logic and mathematical existence, in Krüger and Falkenbburg (1995), pp. 33–82.

Hilbert, D. (1899). Grundlagen der Geometrie. In *Festschrift zur Feier der Enthullung des Gauss-Weber-Denkmals in Gottingen* (1st ed., pp. 1–92). Leipzig: Teubner.

Hilbert, D. (1900a). Mathematische Probleme. *Nachrichten von der Koniglichen Gesellschaft der Wissenschaften zu Gottingen, Math.-Phys. Klasse*, 253–297 (Lecture given at the International Congress of Mathematicians, Paris, 1900. Partial English translation in Ewald (1996), pp. 1096–1105).

Hilbert, D. (1900b). From Mathematical Problems. In W. Ewald (Ed.), From Kant to Hilbert: A Source Book in the Foundations of Mathematics, Oxford: Clarendon Press, pp. 1096–1104.

Hilbert, D. (1905). Ueber die Grundlagen der Logik und der Arithmetik. In A. Krazer (Ed.), *Verhandlungen des dritten Internationalen Mathematiker-Kongresses in Heidelberg vom 8. bis 13. August 1904* (pp. 174–185). Leipzig: Teubner (English translation in van Heijenoort (1967), pp. 129–138).

Hilbert, D. (1918). Axiomatisches Denken. *Mathematische Annalen, 78*, 405–415 (Lecture given at the Swiss Society of Mathematicians, 11 September 1917. English translation in Ewald (1996), pp. 1105–1115).

Hilbert, D. (1920). *Probleme der Mathematischen Logik*. Lecture notes by M. Schönfinkel and P. Bernays, summer session 1920, Mathematical Institute Göttingen.

Hilbert, D., & Bernays, P. (1934). *Grundlagen der Mathematik* (Vol. 1). Berlin: Springer.

Kanamori, A. (2004). Zermelo and set theory. *The Bulletin of Symbolic Logic, 10*, 487–553.

Majer, U. (2006). Hilbert's axiomatic approach to the foundations of science—A failed research program? *Hendricks, 2006*, 155–184.

Moore, G. H. (1982). *Zermelo's axiom of choice: Its origins, development and influence*. New York: Springer.

Moore, G. H. (2002). Hilbert on the infinite: The role of set theory in the evolution of Hilbert's thought. *Historia Mathematica, 29*, 40–64.

Muller, F. A. (2004). The implicit definition of the set-concept. *Synthese, 138*, 417–451.

Neumann, J. von: 1925, Eine Axiomatisierung der Mengenlehre, *Journal für die Mathematik 154*, 219–240, in J. von Neumann, Collected Works I, A. H. Taub (ed.), Pergamon Press, Oxford, 1961, 35–47

Park, W. (2008). Zermelo and axiomatic method. *Korean Journal of Logic, 11*, 2-1-57 (in Korean).

Peckhaus, V. (1994). Hilbert's axiomatic programme and philosophy, in E. Knobloch and D. E. Rowe (Eds.) (1994), pp. 91–112.

Peckhaus, V. (2002). Regressive analysis. *Logical Analysis and History of Philosophy, 5*. Available at http://www.uni-paderborn.de/.

Peckhaus, V. (2005). Pro and Contra Hilbert: Zermelo's set theories. *Philosophical Insight into Logics and Mathematics* (Paris: Kime), 199–216. Available at http://www.uni-paderborn.de/.

Poincare. (1909). La logique de l'infini. *Revue de métaphysique et de morale, 17*, 461–482.

Resnik, M. (1980). Frege and the Philosophy of Mathematics, Ithaca and London: Cornell University Press.

Rowe, D. (2000). The calm before the storm: Hilbert's early views on foundations, in F. Hendricks et al. (Eds.), *Proof Theory*, pp. 55–93.

Scott, D. (1974). Axiomatizing set theory. *Axiomatic Set Theory: Proceedings of Symposium in Pure Mathematics, 13*, 207–214.

Shapiro, S. (1997). *Philosophy of mathematics: Structure and ontology*. New York: Oxford University Press.

Stöltzner. (2002). *How metaphysical is deepening the foundations? Hahn and Frank on Hilbert's axiomatic method* (Heidelberger and Stadler, Eds., (2002), pp. 245–262).

Taylor, R. G. (1993). Zermelo, reductionism, and the philosophy of mathematics. *Notre Dame Journal of Formal Logic, 34*(4), 539–563.

Zermelo, E. (1908a). A new proof of the possibility of a well-ordering, pp. 183–198, in *From Frege to Goedel: A source book in mathematical logic, 1879-1931*, edited by Jean van Heijenoort. Cambridge: Harvard University Press, 1967.

Zermelo, E. (1908b). Investigations in the foundations of set theory, pp. 199–215, in *From Frege to Goedel: A source book in mathematical logic, 1879-1931*, edited by Jean van Heijenoort, Cambridge: Harvard University Press, 1967.

Zermelo, E. (1909). Sur les emsembles finis et le principe de l'induction complète. *Acta Mathematica, 32*, 185–193.

Zermelo, E. (1930). Ueber Grenzzahlen und Mengenbereiche: Neue Untersuchungen ueber die Mengenlehre. *Fundamenta Mathematicae, 16*, 29–47.

Chapter 6
Friedman on Implicit Definition: In Search of the Hilbertian Heritage in Philosophy of Science

Abstract Michael Friedman's project both historically and systematically testifies to the importance of the relativized a priori. The importance of implicit definitions clearly emerges from Schlick's General Theory of Knowledge (Schlick 1918). The main aim of this paper is to show the relationship between both and the relativized a priori through a detailed discussion of Friedman's work. Succeeding with this will amount to a contribution to recent scholarship showing the importance of Hilbert for Logical Empiricism.

Keywords David Hilbert · Implicit definition · Logical empiricism
Michael Friedman · Relativized a priori

1 Introduction

Michael Friedman has been prominent in recent reevaluations of logical positivism. His project both historically and systematically testifies to the importance of the relativized a priori. The importance of implicit definitions clearly emerges from Schlick's General Theory of Knowledge. The main aim of this paper is to show the relationship between both and the relativized a priori through a detailed discussion of Friedman's work. Succeeding with this will amount to a contribution to recent scholarship showing the importance of Hilbert for Logical Empiricism. Friedman details how the relativized a priori was viewed by the leading logical positivists, Reichenbach, Schlick, and Carnap. Curiously enough, however, Friedman is rather reticent in presenting his own views about the Hilbertian heritage of logical positivism. To what extent and in what respects were logical positivists indebted to Hilbert? In particular, what exactly did they learn from the notion of implicit definition as the core of Hilbert's axiomatic method? To what extent does the history of implicit definitions overlap with that of the relativizeda priori? To the best of my knowledge, Friedman never made any sustained attempt to answer these

This chapter was originally published as Park (2012).

© Springer International Publishing AG, part of Springer Nature 2018
W. Park, *Philosophy's Loss of Logic to Mathematics*, Studies in Applied Philosophy, Epistemology and Rational Ethics 43, https://doi.org/10.1007/978-3-319-95147-8_6

questions directly. This is quite strange in view of the fact that Friedman frequently invokes the issue of implicit definition in some crucially important contexts throughout his extensive writings on logical positivism.[1]

Paradoxically, Friedman apparently never attempted a full exposition of the relationship between Hilbert and logical positivism, though that relationship is the crux of the matter in understanding both. It is understandable in some sense, for there are many obstacles to overcome before directly confronting this issue. Above all, the idea of implicit definition has never been clarified completely (See Müller 2004). We must wait for a resolution of the famous Frege-Hilbert controversy (See Frege 1980). As a consequence, for too long we have been lacking a sound understanding of the role and function of definitions in mathematics. Further, as evident in Ulrich Majer's work, we still do not quite understand either the continuity or discontinuity of the traditional and Hilbertian axiomatic method. We even fail to decide whether the problem of implicit definition is merely a side issue, as Majer claims.[2] Finally, Carnap and other logical positivists did not explicitly acknowledge their particular debt to Hilbert.[3] Nor did they make their stance against Hilbert and his followers clear enough. However, time ripens for Friedman to do this long awaited homework. We find evidence of this in his most recent studies on Carnap's logic of science (Friedman 2008, 2011). For the problem of implicit definition looms large in Carnap's views on the theoretical terms in science.[4]

My strategy for understanding the relationship between Hilbert and logical positivism is as follows. First, in Sect. 2, I shall try to fathom Friedman's mind about the relationship between Hilbert and logical positivism by probing exactly why Friedman takes issue with implicit definition in the context where he discusses the relativized a priori. In so far as we find the most sophisticated conception of the relativized a priori in later Carnap, as Friedman claims (Friedman 1992, p. 55), the Hilbertian notion of implicit definitions is at least a sine qua non of Carnap's notion of the relativized a priori. Second, in Sect. 3, I shall try to examine whether Carnap simply extends Hilbert's axiomatic method, as Bernays claims (Bernays 1961, 186),

---

[1]Friedman (1999) discussed issues related to implicit definitions or to Hilbert in virtually all chapters. However, the most pertinent one seems to be Chap. 3, "Geometry, Convention, and the Relativized A Priori", which was originally published as Friedman (1994), rather than the chapters dealing with Carnap's views on logico-mathematical truth. Even more important for my purpose is Friedman (1992). Friedman (2002) contains extremely useful information, especially in the endnotes.

[2]See Majer (2002, pp. 214, 219, 223). However, I fail to find any argument in Majer's writings for his claim that implicit definition is merely a side issue.

[3]Majer raises the question "Why didn't the logical empiricists pay more attention to Hilbert and his axiomatic point of view?" (Majer 2002, p. 213).

[4]There are also similar signs from other philosophers that indicate the urgency and timeliness of addressing this issue. For example, Yemima Ben-Menahem devoted a chapter on implicit definition in her recent book Conventionalism (Ben-Menahem 2006). William Demopoulos and Sathis Psillos write extensively on Carnap's views on theoretical terms in science (Demopoulos 2003, 2007; Psillos 1999; Psillos and Christopoulou 2009).

or achieves his own revised and improved version of the axiomatic method, as Friedman claims (Friedman 1992, p. 55). For this purpose, largely based on Majer's and Stöltzner's studies, I shall briefly review Hilbert's discussion of the axiomatic method as the universal method of deepening the foundations of all sciences (Sect. 3.1). Then, I shall discuss Friedman's treatment of Carnap's later study of theoretical terms in science (Sect. 3.2).

2 Implicit Definition and the Relativized a Priori

There is no doubt that implicit definition is an extremely important issue for Friedman, as he discusses implicit definitions in the context where he introduces the relativized a priori. For example, Friedman (1992) writes:

> This new conception of the physical a priori—which we might call the relativized a priori—has thus been illustrated intuitively by the evolution of physical dynamics from Newtonian physics through the general theory of relativity. But how is it more precisely to be understood? How are we to articulate the general conception of principles that are constitutive in Reichenbach's sense in a philosophically satisfactory manner? One important line of thought—developed especially by Schlick in his General Theory of Knowledge—appeals to Hilbert's logically rigorous axiomatization of geometry and the idea of implicit definitions. (Friedman 1992, p. 51)[5]

Here Friedman's choice of the phrase "one important line of thought" must be prudent. In view of the fact that he is discussing the fortuna of relativized a priori in Reichenbach, Schlick, and Carnap, however, "the only one important line of thought" could have served the same purpose. For no alternative way of articulating the relativized a priori other than by implicit definitions seems to be discussed in Friedman (1992).

How could implicit definitions be instrumental in articulating the relativized a priori? According to Friedman, "Hilbertian implicit definitions seem well suited indeed to capture the new conception of the relativized a priori", because from the idea of implicit definitions it follows,

> first, that the special a priori status of geometry is merely a consequence of this definitional role of the axioms, and, second, that what is thus a priori is also merely conventional. (Friedman 1992, p. 51)

---

[5]Friedman (1992) is not included in Friedman's masterpiece on logical positivism, i.e., Friedman (1999), probably because of the overlaps (especially with the final chapter on Carnap's views on analytic truth). Further it is not found in the bibliography either. From my point of view, the exclusion of Friedman (1992) is rather unfortunate, as it seems the most extensive discussion of the problem of implicit definition in logical positivism.

We'll need to discuss the problem of the conventionality of implicit definitions more carefully in due course, but, as a first approximation, Friedman's account of the strong connection of the relativized a priori and implicit definitions seems to be remarkable.

As pointed out by Friedman, the logical positivists begin their philosophizing by rejecting the Kantian analysis of scientific knowledge as consisting of the pure part and the empirical part. Especially, they reject Kant's idea that the pure part consists exclusively of synthetic a priori judgments. Again, as beautifully portrayed by Friedman, they are obviously indebted to nineteenth-century work on the foundations of geometry, which culminates in Einstein's theory of relativity. Interestingly, however, the logical positivists' view of geometry and scientific knowledge is not an empiricist one. It is "neither strictly Kantian nor strictly empiricist" (Friedman 1994, pp. 21–22; 1999, pp. 59–60). Friedman praises Reichenbach for "a particularly striking version of this radically new conception of scientific knowledge" (Friedman 1994, pp. 22–23; 1999, pp. 60–61). And it must be Reichenbach's distinction between axioms of coordination and axioms of connection, and his understanding of the former as constitutive a priori that elicits such a praise.[6]

Unfortunately, according to Friedman, Reichenbach's earlier conception of the relativized a priori is lost due to Schlick's criticism (Friedman 1994, p. 26; 1999, p. 64). Nevertheless, contrary to Reichenbach's initial accusation, Schlick wholeheartedly "accepts the distinction between constitutive principles and empirical laws", Friedman points out:

> …Schlick prefers neither to characterize constitutive principles as a priori nor to conceive the distinction in question as in any way Kantian. (Friedman 1994, p. 25; 1999, p. 63)

According to Friedman, "Schlick chides Reichenbach for neglecting Poincaré", and suggests characterizing constitutive principles "as conventions in the sense of Henri Poincaré" (Friedman 1994, p. 25; 1999, p. 63). And Friedman seems to be right in observing that "Reichenbach adopts Poincaré's terminology of 'convention' from 1922 onward-notably, in Reichenbach (1924/1969 and 1928/1958)" (Friedman 1994, p. 26; 1999, p. 64). The problem is, as Friedman claims, "when Reichenbach acquiesces in the Schlick–Poincaré terminology", "the most important element in his own earlier conception of the relativized a priori is actually lost" (Friedman 1994, p. 26; 1999, p. 64).

Finally, Friedman claims that "in Carnap's Logical Syntax of Language we find a revival of the relativized a priori in something very like Reichenbach's original sense" (Friedman 1994, p. 30; 1999, p. 68). He even suggests that "Carnap's L-rules or analytic sentences can be profitably viewed as a precise explication of Reichenbach's notion of the relativized a priori" (Friedman 1994, p. 31; 1999, p. 69).

---

[6]"The former are constitutive a priori in that, first, they are not themselves subject to straightforward empirical confirmation or disconfirmation by measuring parameters and instantiating laws, and second, they first make possible the confirmation and disconfirmation of empirical laws properly so-called (viz., the axioms of connection)". (Friedman 1994, p. 23; 1999, p. 61); Cf. Reichenbach (1920/1965).

We have seen above how Friedman outlines the fortuna of the relativized a priori in the hands of the leading logical positivists, Reichenbach, Schlick, and Carnap. For our present purpose, however, we need to know what exactly the role and function of implicit definitions is in the history of the relativized a priori. In other words, we need to check to what extent the history of implicit definitions overlaps with that of relativized a priori. In Friedman (1994, 1999), one might get the impression that Schlick's acceptance of the Hilbertian idea of implicit definition is at least partially to blame for the loss of the relativized a priori for a long time. For example, Friedman writes:

> In sum, Schlick's conception of the peculiarly nonempirical status of physical geometry rests on very general logical and epistemological doctrines: on Hilbert's doctrine of implicit definitions combined with what we now call Duhemian holism .... Reichenbach's account of physical geometry (1920/1965), on the other hand, does not depend on general logical and epistemological doctrines of this kind (although Reichenbach does accept Hilbertian implicit definitions as an account of pure mathematics). (Friedman 1994, p. 27; 1999, p. 65; See also Friedman 1994, p. 26 and p. 32; 1999, p. 64 and p. 70)

If such an impression is not entirely mistaken, then Schlick's understanding of constitutive principles in terms of implicit definitions would be responsible for the loss of the relativized a priori between 1922 and 1934. Also, the relativized a priori revived by Carnap (1937) would have nothing to do with implicit definitions.

However, from Friedman (1992), we can get an entirely different impression about how Friedman views the role and function of implicit definitions in the history of the relativized a priori. For example, Friedman claims that

> [i]t is important, at this point, to note the precise sense in which Carnap has radically revised and improved the original Hilbertian notion of implicit definition. (Friedman 1992, p. 55)

No matter how "radically revised and improved", Carnap's notion of implicit definitions, and eo ipso his notion of the relativized a priori, must have something to do with the Hilbertian notion of implicit definitions. In other words, the Hilbertian notion of implicit definition cannot simply play a negative role and function in the history of the relativized a priori. At the same time, we cannot just count Schlick's acceptance of implicit definitions as responsible for the loss of the relativized a priori. It is accepted by Reichenbach (at least in pure mathematics) as well as by Carnap (at least as in revised and improved form). In fact, Friedman (1992) provides us with ample evidence that he grants some quite positive roles and functions to the Hilbertian notion of implicit definitions in the history of the relativized a priori.

Before turning to some such evidence, however, let us see first how Friedman reports about two overwhelming problems of understanding of the relativized a priori in terms of implicit definitions, which he calls "the definitional—and hence analytic—a priori":

> In the first place, it does not harmonize at all well with the analytic a priori stemming from Frege's logicism. (Friedman 1992, pp. 51–52)

> ...

> In the second place, however, the notion of implicit definition is far too weak satisfactorily
> to ground the relativized or physical a priori as such—even if we ignore all questions about
> harmonizing it with the absolute or logical a priori. (Friedman 1992, p. 52)

In order to elaborate the first problem, Friedman points out that "implicit definitions alone are quite insufficient to demonstrate the analyticity of a particular formalized theory" (Friedman 1992, p. 52). On the other hand, the second problemarises, since "no distinction within a given axiomatized theory (between axioms of coordination and axioms of connection, between convention and fact) is generated by providing a system of implicit definitions." (Friedman 1992, p. 52).[7]

Now, immediately after pinning down two problems of implicit definitions, Friedman presents an extremely suggestive and perceptive paragraph, which I would like to quote in full:

> Now Carnap, as a good logicist and student of Frege, was never satisfied with implicit
> definition as an account of definitional or analytic truth. Unlike Frege himself, however,
> Carnap also wanted somehow to accommodate the new relativized or physical a priori
> arising in the context of recent work in the foundations of geometry and relativity theory.
> Carnap's problem here is therefore to harmonize the absolute logical a priori of Frege's new
> logic (which articulates the most general conditions of objective judgement and truth) with
> the relativized physical a priori of the new geometry and physics (which is to delineate the
> conventional or definitional parts of particular scientific theories). And it is precisely here,
> I believe, that Carnap's unique contribution to the theory of a priori is to be found. In
> contradistinction to the purely "conventionalist" efforts of Schlick and Reichenbach,
> Carnap wanted to incorporate the logicist theory of analytic truth; in contradistinction of the
> logicism of Frege, Russell, and early Wittgenstein, he also wanted to find a place for the
> relativized physical a priori and thus for conventionalism. (Friedman 1992, pp. 52–53)
> [Italics are mine.]

Here Friedman is trying to do justice to both Fregean and Hilbertian heritage in Carnap's project. However, unlike in Frege's case, Hilbert is never mentioned explicitly. As a consequence, one may not fully appreciate Carnap's debt to Hilbert. So, it might be useful to highlight some of the positive influence of Hilbert and especially the Hilbertian notion of implicit definitions on Carnap conceded in this paragraph: (1) "Carnap also wanted somehow to accommodate the new relativized or physical a priori arising in the context of recent work in the foundations of geometry and relativity theory"[8]; (2) Carnap's problem is to harmonize the absolute logical a priori of Frege's new logic with the relativized physical a priori of the new

---

[7]Elsewhere Friedman explains that the abstract mathematical representations lying at the basis of Einstein's general theory of relativity made both logical positivists and Einstein discern an intimate relationship between the theory of relativity and Hilbert's axiomatic conception of geometry. He points out that "[o]n this new view of mathematics there is thus more need than ever for principles of coordination to mediate between abstract mathematical structures and concrete physical phenomena" (Friedman 2001, p. 71).

[8]This is the italicized part in the paragraph I've just quoted from Friedman (1992, pp. 52–53). By "recent work in the foundations of geometry and relativity theory", he seems to have Reichenbach in mind, but in order to have a larger perspective, one may also include the late nineteenth century luminaries such as Helmholtz, Poincaré, and Hilbert as well as Einstein, Reichenbach, and Schlick.

geometry and physics (which is to delineate the conventional or definitional parts of particular scientific theories); (3) Carnap "wanted to find a place for the relativized physical a priori and thus for conventionalism". As we saw above, Hilbert's work must be one of the most distinguished achievements in the context of recent work in the foundations of geometry and relativity theory. Hilbert also contributes most to delineating the conventional or definitional parts of particular scientific theories.

Probably, Friedman has to justify such an understanding of the positive influence of Hilbert on Carnap against the contrary evidence that Carnap rejected the Hilbertian implicit definitions. At the beginning of the paragraph quoted above, Friedman already concedes that "Carnap was never satisfied with implicit definition as an account of definitional or analytic truth". He is also fully conscious of the fact that Carnap had earlier rejected implicit definitions as proper definitions in "Eigentliche und uneigentliche Begriffe" (Symposion 1927) (Friedman 1992, p. 53, n. 8; Here Friedman refers to Carnap 1927). Nevertheless, Friedman believes that Carnap tried to accommodate the Hilbertian notion of implicit definitions, so he is slowly but surely insinuating his belief by enumerating the evidence. Here are some of them:

> In the Aufbau, on the other hand, although Carnap officially rejects implicit definitions as proper definitions, he nonetheless employs a method of definition that is very close indeed to that of implicit definition: in effect, he develops his "purely structural definite descriptions" by adding an empirical existence and uniqueness condition to a standard implicit definition. (Friedman 1992, p. 53; See Section 15 of the Aufbau.)

> Carnap is able coherently to reconcile the two strands of his thought only in Logical Syntax of Language, where he first develops his mature conception of analytic truth. And it is noteworthy that he explicitly presents this conception as a reconciliation of Fregean and Hilbertian ideas (Section 84 of the Aufbau). (Friedman 1992, p. 54)

Friedman's interpretation of Carnap's project in Logical Syntax of Language as a reconciliation of Fregean and Hilbertian ideas culminates in his pursuit of "the precise sense in which Carnap has radically revised and improved the original Hilbertian notion of implicit definition":

> Like Hilbert, Carnap conceives the primitive terms of a given formal language as having no meaning whatsoever outside of the language itself: their meanings, that is, are entirely specified by the rules of the language. Moreover, the rules of a given language are constitutive of the logic of that language and, accordingly, of the notions of correctness and incorrectness relative to that language. The key point of difference, however, is that all rules of the language are not on the same level in this regard: the logic of the language is constituted by the analytic sentences or logical rules, and by these rules alone; the synthetic sentences or physical rules are not constitutive of the language in question in this way, but rather make factual and empirical assertions about an already so-constituted subject matter. And it is by thus incorporating a sharp distinction, within a given formalized language, between two essentially different components, that Carnap's conception of analytic truth overcomes the decisive objection to implicit definitions discussed above (according to which the latter induces no such distinction). (Friedman 1992, pp. 55–56)

Interestingly, however, it was already pointed out by Friedman (in a paragraph immediately preceding the quoted one) that in Carnap's physical language we have both analytic and synthetic sentences with sharp distinction between them. Moreover, there Friedman made an extremely suggestive remark that "[i]t is in thus bringing together pure and applied mathematics that the demands of logicism and formalism are simultaneously met" (Friedman 1992, p. 55).[9]

3 Carnap's Debt to Hilbert

We saw above how Friedman views the role and function of the Hilbertian notion of implicit definitions in the history of the relativized a priori. It turns out that it does not have only the negative role of hindering the continuous development of the relativized a priori by its connection with conventionalism. If Friedman is right, the evolution of implicit definition from Hilbert to Carnap itself is a fundamental part of the history of the relativized a priori. Thus, one might wonder whether Friedman indeed satisfactorily answers my initial question, i.e., "To what extent are the logical positivists indebted to Hilbert?" But the situation is not so simple. Above all, it seems that Friedman has not paid due attention to Hilbert's own notion of implicit definitions in his own writings. As a consequence, we have good reason to believe that Friedman has not fully answered the following questions: Within Hilbert's lifelong project of the axiomatic method, what exactly is the role and function of implicit definitions? How do (or should) we use implicit definitions in "deepening the foundations"? How does Hilbert himself exploit implicit definitions in his own work on the foundations of physics? Without answering these questions, any attempt to appreciate the logical positivists' debt to Hilbert would be incomplete.

Obviously there are several overwhelming obstacles in understanding both the original Hilbertian axiomatic method and Carnap's own (revised and improved?) version of the axiomatic method. Fortunately, we can witness some significant ongoing projects in this regard, which can conveniently be classified into two general categories: (1) "Understanding the Hilbertian Deepening the Foundations", and (2) "Later Carnap's Study of Theoretical Terms in Science". The best examples of (1) can be found in Ulrich Majer's and Michael Stöltzner's studies (See Majer 1996, 2001, 2002, 2006 and Stöltzner 2002, 2003). On the other hand, the best examples of (2) can be found in William Demopoulos', Stathis Psillos', and Friedman's work (See Demopoulos 2003, 2007, Psillos 1999, 2000, and Friedman 2008, 2011). What we need to do is, based on the results in (1) and (2), to determine whether Hilbert's idea of deepening the foundations anticipated Carnap's project of studying theoretical terms in terms of implicit definitions. In other words, we want

---

[9]It is far beyond the scope of this paper to examine whether Friedman is correct here. Though it is highly controversial, to say the least, his suggestion deserves careful attention. For it seems to show the way to an in-depth understanding of the motivations of Carnap's projects in both *Aufbau* and the *Logical Syntax*.

to know whether later Carnap was merely extending Hilbert's axiomatic method or revised and improved it in such a way that we can meaningfully discuss his own version of the axiomatic method.[10]

3.1   Hilbert on Deepening the Foundations[11]

Hilbert's "Axiomatisches Denken" (1918) emphasizes the central role of mathematics among the sciences. There we find the following suggestive claims:

> The procedure of the axiomatic method, as it is expressed here, amounts to a deepening of the foundations of the individual domains of knowledge—a deepening that is necessary for every edifice that one wishes to expand and to build higher while preserving its stability. (Hilbert 1918, p. 1109)

This sentence, which seems to claim that deepening the foundations of the individual science necessarily demands its expansion to its neighbouring fields of knowledge, provides us with an extremely interesting issue.

In fact, the reexamination of Hilbert's views on the foundations of physics is quite in vogue in recent researches on Hilbert. We can only welcome this trend in view of the well-known fact that the problem of axiomatizing physics is the sixth of the twenty-three problems presented by Hilbert in 1900 (See Hilbert 1900). For example, Majer's discussion contains the potentiality of serving as the springboard for further fruitful research:

> ...the application of the axiomatic method in physics is only legitimate insofar as one takes it as an expansion (or transfer) of the axiomatic point of view in geometry. This does not imply that they have to be exactly the same; there may be certain differences due to the differences between geometry and physics. On the other hand, Hilbert was deeply convinced (from the very beginning of his occupation with geometry) that geometry is a natural science, indeed the most fundamental of all natural sciences, and this is the deeper reason why the axiomatic point of view should be transferable to physics. (Majer 2006, p. 164. Majer's emphasis)

Considering Hilbert (1918), I would now like to confirm the continuity of the Euclidean axiomatic approach and Hilbert's axiomatic method by highlighting selectively what is implied by Majer's observations. At the same time, I will probe

---

[10]The former understanding can be found in Bernays (1961): "The general tendency of the Logical Syntax can be said to be an extension of the approach of Hilbert's proof theory. For Hilbert the method of formalization is applied only to mathematics. However, in his lecture "Axiomatic Thought" Hilbert also said: "Everything at all that can be the object of scientific thinking falls under the axiomatic method, and thereby indirectly under mathematics, when it becomes mature enough to form into theory". Carnap goes a step further in this direction in the Logical Syntax, by considering science as a whole as an axiomatic deductive system which becomes a mathematical object through formalization: the syntax of the language of science is metamathematics that is directed towards this object" (Bernays 1961, p. 186).

[11]The basic ideas in Sect. 3.1 were first presented in Park (2008).

the question as to what differences the one and the same axiomatic method would make depending on the peculiarity of the individual science that is to be axiomatized. Hopefully, we may thereby fathom the desiderata for axiomatization of any given field of knowledge.

Immediately before the sentences quoted above, Hilbert explained by exploiting ample examples how the axiomatic method is applied and how the deepening of the foundations results in a given field of knowledge. Let us follow Hilbert's lead to capture the major stages in a schematic fashion.

(1) Collected facts in a given field (of knowledge) can be ordered by a certain conceptual framework.
(2) In establishing the conceptual framework for the field (of knowledge), there are some underlying salient propositions. (The propositions are themselves good enough for building the entire framework according to logical principles.)
(3) The fundamental propositions can be considered as axioms for the individual field (of knowledge). (The development of the field depends only on the additional logical establishment of the conceptual framework.)
(4) The problem of establishing the foundations of the field (of knowledge) has been solved.
(5) However, the solution is only temporary. A need arises to ground the propositions themselves that are fundamental and axiomatic in the field (of knowledge) (Based on Hilbert 1918, pp. 1107–1109; See Park 2008, pp. 19–20).

It is not so difficult to follow Hilbert's reasoning here, but the problem is whether it is also easy to understand the following emergent thought. How to ground those fundamental propositions themselves? Hilbert's answer is as follows:

> So one acquired 'proofs' of the linearity of the equation of the plane and the orthogonality of the transformation expressing a movement, of the laws of arithmetical calculation, of the parallelogram of forces, of the Lagrangian equations of motion, of Kirchhoff's law regarding emission and absorption, of the law of entropy, and of the proposition concerning the existence of roots of an equation. (Hilbert 1918, p. 1109, [7].)

However, Hilbert is quick to point out that those proofs are not in themselves proofs.

> But critical examination of these 'proofs' shows that they are not in themselves proofs, but basically only make it possible to trace things back to certain deeper propositions, which in turn are now to be regarded as new axioms instead of the propositions to be proved. The actual so-called axioms of geometry, arithmetic, statics, mechanics, radiation theory, or thermodynamics arose in this way. These axioms form a layer of axioms which lies deeper than the axiom-layer given by the recently-mentioned fundamental theorems of the individual field of knowledge. (Hilbert 1918, p. 1109, [8].)

If we add Hilbert's explanation like this to the scheme discussed above, then we can have the following stages:

(6) We try to trace those fundamental propositions to certain deeper propositions (by proofs that are not proofs in a rigorous sense).

(7) Those deeper propositions are now regarded as new axioms.

(8) Those new axioms form a layer of axioms which lies deeper than the axiom-layer given by the recently-mentioned fundamental theorems of the individual field of knowledge (Based on Hilbert 1918, p. 1109; See Park 2008, p. 21).

Stöltzner (2002) is extremely helpful in understanding both Hilbert's own notion of deepening the foundations embedded in his own practice and its reception by some of the logical positivists, especially Hahn and Frank. As for the first, he finds at least eight examples of deepening the foundations in Hilbert. As for the second, he explains why some of the examples of the Hilbertian deepening the foundations are thought to be metaphysical to Hahn and Frank. It is simply impossible for me to go into any of these two crucial issues. Let me just quote Stöltzner's conclusion, which I agree with wholeheartedly:

> Summing up, while those types of "deepening the foundations" which respected the borderline between mathematics and empirical science were acceptable for Logical Empiricists, those which do not, sometimes contained important breakthroughs and sometimes a priori prejudices. Since Hahn and Frank wanted to make sure that such errors could never occur in a methodologically sober science, their containment strategy against metaphysics had to pay the price of simply ignoring many positive aspects of Hilbert's axiomatic method. In view of the subsequent rise of mathematical physics this was not an entirely fortunate strategy and so the problems centering around Hilbert's "deepening the foundations" are still with today's philosophy of science. (Stöltzner 2002, p. 259; See also Stöltzner 2003)

It is quite natural to be curious about the responses of other logical positivists, especially Reichenbach, Schlick, and Carnap.[12] Do they share Hahn's and Frank's worries about the metaphysical dangers of "Tieferlegung" and the non-Leibnizian pre-established harmony?[13] It seems to me that Carnap would not find any problems with either the Hilbertian notion of deepening the foundation or the (non-Leibnizian) pre-established harmony of mathematics and physics. However, in view of both his anti-metaphysical attitude and other logical positivists' worries about the metaphysical implications of Hilbertian deepening the foundations, he was bound to be prudent. In the next section, we may gather some further clues as to how Carnap tried to appropriate Hilbert's axiomatic method without becoming too metaphysical.

---

[12]Reichenbach's case is discussed in Stöltzner (2003), which deals with the problem of the principle of least action.

[13]This question is raised because to Hahn and Frank Hilbert's allusion to the (non-Leibnizian) preestablished harmony between our thought and the course of the world (or between mathematics and physics) was "thoroughly mysterious". As what Stöltzner counts as "the main charge against Hilbert", Hahn wrote: "Why should what is compelling to our thought also be compelling to the course of the world? Our only recourse would be to believe in a miraculous pre-established harmony between the course of our thought and the course of the world, an idea which is deeply mystical and ultimately theological" (Hahn 1987, p. 28; Stöltzner 2002, p. 251).

3.2 Friedman on Carnap's Later Views on Theoretical Terms in Science

Friedman (2001) points out that "[i]n his Logical Syntax of Language of 1934 Carnap urged that we should extend Hilbert's method from logic to the whole of philosophy" (Friedman 2001, p. 16). As Friedman reports, according to Carnap, "Scientific philosophy should now become Wissenschaftslogik—the meta-logical investigation of the logical structures and relations of the total language of science" (Friedman 2001, p. 16). Friedman's most recent studies (Friedman 2008, 2011) of Carnap's later views on the theoretical terms in science, which must be the centerpiece of his Wissenschaftslogik, seem to indicate that Carnap's debt to Hilbert is even more far-reaching. In both articles, Friedman notes that as early as 1939 Carnap shows a keen interest in the "increasing use of the (Hilbertian) axiomatic method in modern physics (1939, p. 209)". (Friedman 2008, p. 395; 2011, p. 253) based on the following text:

> The development of physics in recent centuries, and especially in the past few decades, has more and more led to that method in the construction, testing, and application of physical theories which we call formalization, i.e., the construction of a calculus supplemented by a [partial—MF] interpretation. It was the progress of knowledge and the particular structure of the subject matter that suggested and made practically possible this increasing formalization. In consequence it became more and more possible to forego an "intuitive understanding" of the abstract terms and axioms and theorems formulated with their help. (Carnap 1939, p. 209; quoted in Friedman 2008, p. 395; 2011, p. 253)

Again, in both articles, Friedman makes the same comment:

> Carnap sees the theories of relativity and quantum mechanics as the culmination of this development—where the use of highly abstract terms introduced by something like Hilbertian implicit definitions (terms such as 'electron', 'electromagnetic field', 'metric-tensor', 'psi-function', and so on) has become a pervasive and essential feature of physical practice". (Friedman 2008, p. 396; 2011, p. 253)[14]

Referring to Carnap (1956), Friedman explains that "only the observational terms are semantically interpreted", while "the theoretical terms are only implicitly defined" (Friedman 2011, p. 253). Only through "correspondence rules, which set up (lawlike) relationships among theoretical and observational terms", the theoretical terms "receive a partial interpretation in terms of the connections they induce among observables" (Friedman 2011, p. 253).

In Carnap's later studies on theoretical terms in science, we find brilliant discussions of Ramsey sentences and Carnap sentences. Carnap (1959/2000) spiritedly demonstrates how excited he himself was:

---

[14]Also, in both articles, Friedman emphasizes the fact that as late as Carnap (1966/1974) Carnap finds in the modern axiomatic method "our very best hope for progress" (Friedman 2008, pp. 398–399; 2011, p. 261).

Well, we could do it if we found a way of giving explicit definitions for all the theoretical terms in the observation language. And this is the question which I want to raise now: is that possible? I thought very briefly about that question years ago and I just dismissed it from my mind, because it seems so obvious that it is impossible. Everybody knows that the theoretical terms are introduced by postulates just because we cannot give explicit definitions of them on the basis of the observational terms alone, even if we add a strong logic. At least, that seemed to be the case and therefore I did not think more about it, although, if we could do it, that would be a great advantage. Now, it is possible. I found that only a few weeks ago and I hope I have not made a mistake ... So, in the hope that there is something in it, I will now present the way of doing this by explicit definitions, which is really so surprising that I still can hardly believe it myself. (Carnap 1959/2000, p. 168)

As we saw above, Friedman finds a serious problem with the notion of implicit definition in that it does not secure a sharp distinction between analytic and synthetic sentences. We also noted that Friedman views later Carnap as trying to reconcile Frege's and Hilbert's views about implicit definitions. Now, in Carnap (1959/2000), we seem to have the culmination of Carnap's lasting effort. Finally, he achieves what seems impossible.

Even if Carnap achieves the ultimate reconciliation of Frege and Hilbert, it is still unclear whether it is the revised and improved (his own version of) axiomatic method or just an extension of the Hilbertian axiomatic method. In other words, there still is room for asking whether Hilbert's axiomatization of physical theories can be an anticipation of Carnap's views on the theoretical terms in science. (See Carnap 1961 for his adoption of the Hilbertian e-operator in this regard.)

4 Concluding Remarks

I hope that what I have pursued in this article has at least some important implications for Hilbertian as well as Carnapean scholarship. Above all, it may show us how to find a way out from what I call the Hilbertian dilemma: Are we to emphasize the novelty of Hilbert's axiomatic method or the continuity of the history of axiomatic method? If we grasp the second horn, as Majer does, we might be giving up the hope of understanding the so-called second birth of mathematics in the nineteenth and twentieth centuries.[15] If we grasp the first horn, as logical positivists and Friedman do, we should make clear how the method of implicit definition truly works in scientific as well as mathematical practice. The significance of the Hilbertian dilemma cannot be too much emphasized, for in a broader context, it challenges us by asking whether the notion of implicit definition as the

---

[15]This way of setting up a dilemma is, of course, controversial. For example, Majer might object that there is no problem for Hilbertians grasping the second horn of the alleged dilemma. For they could understand Hilbert's axiomatic as achieving a real progress over the old Euclidean axiomatic. However, in so far as there is an urgent need to distinguish sharply between the formalist position and the original Hilbertian axiomatic method, the threat of the dilemma seems genuine.

core of Hilbert's axiomatic method is a revolutionary change in the classical model of science stemming from Aristotle.

After having examined Friedman's discussion of the fortuna of implicit definitions in Hilbert and Carnap, we seem to face another dilemma, which may be called the Carnapian dilemma: Are we to emphasize the novelty of Carnap's axiomatic method or the continuity of the history of the axiomatic method? If we grasp the first horn, as Friedman does,[16] we should be able to pin down what exactly are the revised and improved aspects of the Carnapian axiomatic method. If we grasp the second horn, as Bernays does, we might lose the chance of appreciating the truly revolutionary character of Carnapian Wissenschaftslogik.

We seem to have a desperate need to learn from Hilbert and Carnap how not to make philosophy of mathematics and philosophy of science irrelevant to mathematical and scientific practice, because it seems nowadays almost an axiomatic belief among mathematicians, scientists, and engineers that contemporary philosophy (if not traditional philosophy) is irrelevant to their practice. Apart from the Einstein-Hilbert priority debate, Hilbert was a first rate mathematician-scientist. In other words, his philosophy cannot be irrelevant to mathematical and scientific practice. Similar observations can be made for Carnap and other leading members of logical positivism. No matter what problems it might have, it was indeed a great intellectual movement led by scientifically minded philosophers and philosophically minded scientists. Recovering the Hilbertian heritage could be the final magic touch for the recent (seemingly quite successful) re-evaluation of logical positivism.

Acknowledgements I am indebted to the editors and the two anonymous reviewers of Erkenntnis for their extremely helpful criticisms and suggestions. For the last 10 years, Young-Sam Chun, Junyong Park, Wonbae Choi, and Jinhee Lee shared with me enthusiasm for the history of the axiomatic method. I presented an earlier version of this paper at the international conference on "The Future of Philosophy of Science" held at Tilburg University in April 2010. I wish to thank the organizers and the referees of the conference as well as the participants including Michael Stöltzner, Jeongmin Lee, and especially Michael Friedman for their suggestions, advice, encouragement, and thought-provoking questions. Also, many thanks are due to the written comments of Nino B. Cocchiarella, which were truly instrumental at the final stage.

References

Ben-Menahem, Y. (2006). *Conventionalism*. Cambridge: Cambridge University Press.
Bernays, P. (1961). Zur Rolle der Sprache in Erkenntnistheoretischer Hinsicht. *Synthese, 13,* 185–200. English translation in Bernays Project Text No. 25.
Carnap, R. (1927). Eigentliche und uneigentliche Begriffe. *Symposion, 1,* 355–374.
Carnap, R. (1937). *The logical syntax of language*. London: Kegan Paul.

---

[16]Majer might grasp this horn for an entirely different reason, because he believes that Carnap's idea "to generalize the concept of an axiomatic system in such a way that geometry ... turns out to be nothing but a logical theory" is "definitely not Hilbert's point of view" (Majer 2002, p. 220).

Carnap, R. (1939). Foundations of logic and mathematics. In O. Neurath, R. Carnap, & C. Morris (Eds.), *International encyclopedia of unified science* (Vol. 1, pp. 139–213). Chicago: University of Chicago Press.

Carnap, R. (1956). The methodological character of theoretical concepts. In H. Feigl & M. Scriven (Eds.), *Minnesota studies in the philosophy of science* (Vol. I, pp. 38–76)., The foundations of science and the concepts of psychology and psychoanalysis Minneapolis: University of Minnesota Press.

Carnap, R. (1959/2000). In S. Psillos (Ed.), *Theoretical concepts in science* (pp. 158–172).

Carnap, R. (1961). On the use of Hilbert's e-operator in scientific theories. In Y. Bar-Hillel, et al. (Eds.), *Essays on the foundations of mathematics* (pp. 156–164). Jerusalem: The Magnus Press.

Carnap, R. (1966/1974). *An introduction to the philosophy of science*. New York: Basic Books.

Demopoulos, W. (2003). On the rational reconstruction of our theoretical knowledge. *British Journal for the Philosophy of Science, 54*, 371–403.

Demopoulos, W. (2007). Carnap on the rational reconstruction of scientific theories. In M. Friedman & R. Creath (Eds.), *The Cambridge companion to Carnap* (pp. 248–272). Cambridge: Cambridge University Press.

Frege, G. (1980). David Hilbert. In G. Gabriel, et al. (Eds.), *Philosophical and mathematical correspondence* (pp. 31–52). Chicago: The University of Chicago Press.

Friedman, M. (1992). Carnap and a priori truth. In D. Bell & W. Vossenkuhl (Eds.), *Science and subjectivity* (pp. 47–60). Berlin: Akademie.

Friedman, M. (1994). Geometry, convention, and the relativized a priori. In W. Salmon & G. Wolters (Eds.), (pp. 21–34); *Reprinted in Friedman* (1999). (Chap. 3, pp. 59–70).

Friedman, M. (1999). *Reconsidering logical positivism*. Cambridge: Cambridge University Press.

Friedman, M. (2001). *Dynamics of reason*. Stanford, California: CSLI Publications.

Friedman, M. (2002). Geometry as a branch of physics: Background and context for Einstein's 'geometry and experience'. In B. David & D. B. Malament (Eds.), *Reading natural philosophy: Essays in the history of philosophy of science and mathematics* (pp. 193–229). Chicago: Open Court.

Friedman, M. (2008). Wissenschaftslogik: The role of logic in the philosophy of science. *Synthese, 164*, 385–400.

Friedman, M. (2011). Carnap on theoretical terms: Structuralism without metaphysics. *Synthese, 180*, 249–263.

Hahn, H. (1987). Logic, mathematics, and knowledge of nature. In B. McGuinness (Ed.), *Unified science* (pp. 24–45). Dordrecht: Kluwer.

Hilbert, D. (1900). Mathematische Probleme. In *Nachrichten von der Koniglichen Gesellschaft der Wissenschaften zu Gottingen, Math.-Phys.* Klasse (pp. 253–297). Lecture given at the international congress of mathematicians, Paris, 1900. Partial English translation in Ewald (1996). (pp. 1096–1105).

Hilbert, D. (1918). Axiomatisches Denken. *Mathematische Annalen, 78*, 405–415. In Lecture given at the Swiss Society of Mathematicians, September 11, 1917. English translation in Ewald (1996), pp. 1105–1115.

Majer, U. (1996). Hilbert's criticism of Poincaré's conventionalism. In G. Heinzmann, et al. (Eds.), *Henri Poincaré: Science and philosophy* (pp. 355–364). Berlin: Akademie Verlag.

Majer, U. (2001). The axiomatic method and the foundations of science: Historical roots of mathematical physics in Göttingen (1900–1930). In M. Rédei & M. Stöltzner (Eds.), *John von Neumann and the foundations of quantum physics* (pp. 11–33). Boston: Kluwer.

Majer, U. (2002). Hilbert's program to axiomatize physics (in analogy to geometry) and its impact on Schlick, Carnap and other members of the Vienna Circle. In M. Heidelberger & F. Stadler (Eds.), *History of philosophy and science* (pp. 213–224). Boston: Kluwer.

Majer, U. (2006). Hilbert's axiomatic approach to the foundations of science—a failed research program? In V. F. Hendricks (Ed.), *Interactions: Mathematics, physics and philosophy, 1860–1930* (pp. 155–184). Dordrecht: Springer.

Müller, F. A. (2004). The implicit definition of the set-concept. *Synthese, 138*, 417–451.

Park, W. (2008). Zermelo and the axiomatic method. *Korean Journal of Logic, 11*(2), 1–57. (in Korean).

Park, W. (2012). Friedman on implicit definitions: In search of the Hilbertian heritage in philosophy of science. *Erkenntnis, 76*(3), 427–442.

Psillos, S. (1999). *Scientific realism: How science tracks truth*. London and New York: Routledge.

Psillos, S. (Ed.). (2000). Rudolf Carnap's theoretical concepts in science. *Studies in History and Philosophy of Science, 31*, 151–172.

Psillos, S., & Christopoulou, D. (2009). The a priori: Between conventions and implicit definitions. In N. Kompa, et al. (Eds.), *The a priori and its role in philosophy* (pp. 205–220). Paderborn, Germany: Mentis.

Reichenbach, H. (1920/1965). *The theory of relativity and a priori knowledge* (M. Reichenbach, Trans.). Los Angeles: University of California Press.

Reichenbach, H. (1924/1969). *Axiomatization of the theory of relativity* (M. Reichenbach, Trans.). Los Angeles: University of California Press.

Reichenbach, H. (1928/1958). *The philosophy of space and time* (M. Reichenbach & J. Freund, Trans.). New York: Dover.

Schlick, M. (1918/1985). *General theory of knowledge* (A. Blumberg, Trans.). La Salle: Open Court.

Stöltzner, M. (2002). How metaphysical is "deepening the foundations"?—Hahn and Frank on Hilbert's axiomatic method. In M. Heidelberger & F. Stadler (Eds.), *History of philosophy and science* (pp. 245–262). Boston: Kluwer.

Stöltzner, M. (2003). The principle of least action as the logical empiricist's Shibboleth. *Studies in History and Philosophy of Modern Physics, 34*, 285–318.

Chapter 7
Between Bernays and Carnap

Abstract Bernays has not drawn scholarly attention that he deserves. Only quite recently, the reevaluation of his philosophy, including the projects of editing, translating, and reissuing his writings, has just started. As a part of this renaissance of Bernays studies, this chapter tries to distinguish carefully between Hilbert's and Bernays' views regarding the axiomatic method. We shall highlight the fact that Hilbert was so proud of his own axiomatic method on textual evidence. Bernays' estimation of the place of Hilbert's achievements in the history of the axiomatic method will be scrutinized. Encouraged by the fact that there are big differences between the early middle Bernays and the later Bernays in this matter, we shall contrast them vividly. The most salient difference between Hilbert and Bernays will shown to be found in the problem of the uniformity of the axiomatic method. In the same vein, we will discuss the later Bernays' criticism of Carnap, for Carnap's project of philosophy of science in the late 1950s seems to be a continuation and an extension of Hilbert's faith in the uniformity of the axiomatic method.

Keywords Bernays · Axiomatic method · Hilbert · Carnap · Deepening the foundations · Application of mathematics

1 Introduction

There is no doubt that Bernays is one of those who contributed most to mathematical logic, foundations of mathematics, and philosophy of mathematics. Surprisingly, however, he has been usually mentioned as a collaborator of David Hilbert, or as an important set theorist contributing to the so-called NBG set theory,

An earlier version was published in Korean as Park (2011).

which has been discussed extensively as an alternative to the standard ZFC set theory. Only quite recently, his philosophy is becoming to draw renewed attention, which can be attested by the so-called Bernays project.[1] Welcoming this trend, I will attempt to distinguish carefully Bernays' thought from that of Hilbert by focusing on the axiomatic method.

Hilbert was not only a distinguished modern mathematician but also a philosopher (in every sense of the word), who left unmistakable influence on contemporary science and philosophy by his logic and foundations of mathematics. However, he himself always felt that his background in philosophy was somewhat limited. As a consequence, from his early years on he tried to sustain close friendly relationship with philosophers such as Husserl or Nelson (see Reid 1996). Bernays became Hilbert's assistant in 1917.[2] As is well-known, the collaboration of Hilbert and Bernays in the 1920s was a great success in the foundations of mathematics. It culminated in the publication of *Grundlagen der Mathematik* (Vol. 1, 1934; Vol. 2, 1939). Further, in fact, Bernays is known to be the real author of these volumes.[3] Be that as it may, I would like to emphasize the point that the initial reason for Hilbert to recruit Bernays as his assistant was that Bernays was not only an able mathematician and physicist but also a trained philosopher.[4] Thus, it is impossible to discuss the so-called Hilbert program centering around 1917 through 1925 without understanding Bernays' philosophical and methodological views about the axiomatic method.

For convenience, let us distinguish between early Hilbert (until the publication of Foundations of Geometry (1899)), middle Hilbert (1900–1917), and the later Hilbert (after 1917). Most of the previous studies paid less than enough attention to

---

[1]Parsons (2006) is a research done in connection with the publication project of the English translation of Bernays' work. Feferman (2008) can be counted as a study based on the publication of Collected Papers of Gödel, especially his correspondences (Vols. 4 and 5).

[2]Corry claims that the expression "assistant" is somewhat misleading (Corry 2004, p. 70).

[3]For example, see Specker (1979, p. 385), Zach (1999, pp. 344–345). Zach not only claims that Hilbert and Bernays (1934, 1939) was essentially written by Bernays alone, but also that even in their early collaboration (1917–1918) Bernays contributed at least equally as Hilbert.

[4]According to Specker (1979), which is a very brief biography of Bernays, in 1912 Bernays received a doctoral degree for his study on the theory of numbers under the supervision of Landau at Göttingen. And, in the same year he acquired a license for professorship with his study on the theory of function under Zermelo at the University of Zürich. On Hilbert's invitation, he moved to Göttingen, and received once more the license for professorship with his work on logic. But, when he was an undergraduate student at Berlin, he learned physics from Plank at Berlin, and from Born at Göttingen. On the other hand, he learned philosophy from Riehl, Stumpf, and Cassirer at Berlin, and from Nelson at at Göttingen. Under the influence of Lelson, who was one of the center figures of neo-Friesianism, Bernays wrote a paper on moral philosophies of Sidgwick and Kant in the 1910's. Bernays' background in physics, there are useful and important pieces of information in Corry (2004, pp. 295–296). Bernays was deeply involved in mathematical physics and the foundational issues related to it, publishing Bernays (1913). Also, he lectured on gas theory up to 1920 at Göttingen. We cannot simply bypass Corry's observation that there is an interesting parallel between Zermelo and Bernays in that both showed a transition of interest from mathematical physics to foundations of mathematics.

the development of Hilbert's thought, by failing to distinguish between early, middle, and later Hilbert. Also, they failed to report fairly the fact that Hilbert was indebted enormously to his collaborators. Furthermore, they did not distinguish sharply between Hilbert's own thought and the views of his collaborators, thereby failed to comprehend not only the unique views of each of them but also their shared views. These triple failures are intricately intertwined. There is an urgent need to amend these by echoing recent renaissance of Hilbert studies, which can be seen from the publication of the nachlass, correspondence, and lecture notes of Hilbert and his collaborators.[5] At the same time, we need to distinguish between Bernays' thought in the early, middle and later stages. For convenience's sake, we may call the period before Bernays became the assistant of Hilbert as "the early", the period in which he was a collaborator of Hilbert, i.e., between 1917 and 1939 as "the middle', and the period after 1939 as "the later" stages.[6] Also, we need to go one step further by subdividing the middle Bernays as "the early middle Bernays" and "the later middle Bernays". After all, it will become crucial to understand how Hilbert and Bernays interacted in the period of their close collaboration, at least strategically, such a tentative distinction between the different stages in Bernays' development of thought would have some heuristic value. Of course, from that perspective, Gödel's 1931 incompleteness theorem could be an important date of effective watershed within this period.

My strategy is as follows. In Sect. 2, I shall highlight the point that Hilbert was amazingly proud of his own axiomatic method by citing textual evidences. In Sect. 3, I shall review Bernays' views on the place of Hilbert's axiomatic method in the entire history axiomatic method. The most interesting point is that the middle Bernays and the later Bernays show remarkably different positions to this problem. Enlightened by this, in Sect. 3.1, I shall first contrast the views of the early middle Bernays and that of the later Bernays. In Sect. 3.2, I shall argue that, as a consequence of these discussions, the most salient difference between the views of Hilbert and Bernays is found in the issue of the uniformity of the axiomatic method. In Sect. 4, Bernays' criticism of Carnap will be discussed extensively, for Carnap's project in philosophy of science after the mid 1950s seems to succeed the Hilbert's belief in the uniformity of the axiomatic method.

---

[5]I have in mind the Collected works of Gödel, Zermelo, Carnap, Hilbert-Bernays project, and Bernays project.

[6]Parsons (2006) takes 1945 as the starting point for the later Bernays. That seems to be a very safe choice that allows at least 6 years from the period of collaboration with Hilbert. However, such a choice could be a too safe one, since there is virtually an unanimous agreement that the true author of Hilbert and Bernays (1934, 1939) was Bernays. We need to trace the beginning of the period of the later Bernays back to sometime after 1934 when Bernays was expelled to Zürich by Nazis, especially when he was attracted to Ginseth's thought. For the purpose of this paper, this problem may be ignored.

2   Hilbert's Axiomatic Method and Deepening the Foundations

Apparently, Hilbert was so proud of his own axiomatic method. As is well-known, Hilbert is respected as the father of contemporary mathematics, who left his mark virtually all fields of mathematics. There are many trademarks of Hilbert, including "Hilbert's problems", "Hilbert program", "Hilbert's proof theory", and "Hilbert's metamathematics". In view of all this, it is not a small matter that Hilbert himself showed such a pride for his own axiomatic method (more than anything else) throughout his entire career. Why did he sustained such affection and pride for his own axiomatic method?

Evidence is everywhere to testify that Hilbert was so proud of his own axiomatic method. Above all, he contrasted the genetic method traditionally used in the study of the concept of number with his own axiomatic method, and asked whether each of these is the only method appropriate for geometry, and which is more advantageous in the logical study in the fields of physical sciences, including mechanics. And, he explicitly announced his position:

> My opinion is this: *Despite the high pedagogic and heuristic value of the genetic method, for the final presentation and the complete logical grounding [Sicherung] of our knowledge the axiomatic method deserves the first rank*". (Hilbert 1900, pp. 1093, [3]. Hilbert's emphasis)

Hilbert discussed the point that his axiomatic method (as a superior method) can be applied also to the study of the foundations of arithmetic elsewhere. (Hilbert 1905, p. 130) There Hilbert pointed out that, while there is an agreement in the aim and direction of the research on the foundations of geometry, there are extremely diverse and conflicting opinions on the foundations of geometry. In order to efficiently suggest the major difficulties, he contrasted the positions of Helmholtz, Kronecker, and Christoffel as a dogmatist, an empiricist, and an opportunist, respectively. Then, he enumerated Frege, Dedekind, and Cantor as those who did deeper studies on the essence of integers. At the same time, he expressed his complaints to each of them. Only then, he introduced his own axiomatic method as the solution to handle all the difficulties:

> It is my opinion that all the difficulties touched upon can be overcome and that we can provide a rigorous and completely satisfying foundation for the notion of number, and in fact by a method that I would call *axiomatic* and whose fundamental idea I wish to develop briefly in what follows. (Hilbert 1905, p. 131)

In his Zürich lecture, presented immediately after the priority debate with Einstein (see Corry 2004), Hilbert enthusiastically argued for the extended application of the axiomatic method to neighboring sciences, including physics. At the beginning of this lecture, Hilbert made the following remark:

> Just as in the life of nations the individual nation can only thrive when all neighbouring nations are in good health; and just as the interest of states demands, not only that order prevail within every individual state, but also that the relationships of the states among

themselves be in good order; so it is in the life of the sciences. In due recognition of this fact the most important bearers of mathematical thought have always evinced great interest in the laws and the structure of the neighbouring sciences; above all for the benefit of mathematics itself they have always cultivated the relations to the neighbouring sciences, especially to the great empires of physics and epistemology. I believe that the the essence of these relations, and the reason for their fruitfulness, will appear most clearly if I describe for you the general method of research which seem to be coming more and more into its own in modern mathematics; I mean the *axiomatic method*. (Hilbert 1918, p. 1107, [1])

In this lecture, Hilbert equated the procedure, which is well-known as his motto of "deepening the foundations", with his axiomatic method.[7] And he structured his lecture with the two requirements the procedure should satisfy, i.e., independence and consistency, as the two major pillars. Now, we may note that Hilbert again celebrated his own axiomatic method in the context of discussing the problem of consistency. According to Hilbert, if a contradiction appears in physical theories, they are always eliminated by selecting different axioms. However, the situation is entirely different in the areas of pure and theoretical knowledge. As an example, Hilbert invoked the so-called Cantor paradox, and pointed out that because of it mathematicians like Kronecker and Poincaré thought that the existence of set theory, which is the discipline of most fruitful and powerful knowledge, cannot be justified. (Hilbert 1918, pp. 1111–1112, [22–31]) Then, Hilbert triumphantly claims as follows:

> But in this precarious state of affairs as well, the axiomatic method came to the rescue. By setting up appropriate axioms which in a precise way restricted both the arbitrariness of the definitions of sets and the admissibility of statements about their elements, Zermelo succeeded in developing set theory in such a way that the contradictions disappear, but the scope and applicability of set theory remain the same. (Hilbert 1918, pp. 1112, [32])

Furthermore, it is also remarkable that Hilbert never lost that pride over his axiomatic method throughout his life. In Hilbert (1930), which is almost the last major work, Hilbert mentioned an old philosophical problem—"namely, the vexed question about the share which thought, on the one hand, and experience, on the other, have in our knowledge" in the context of enumerating the towering achievements of modern history of science and mathematics. (Hilbert 1930, p. 1157, [1]) What is notable here is that Hilbert must be meditating on the significance of his life-long scientific achievement against such historical background.

According to Hilbert, we can be more hopeful for solving the old problems of philosophy more convincingly and more exactly. (Hilbert 1930, pp. 1157–1158, [2]) And, he cites the two reasons for such optimism. The first reason, which anyone can agree with, is that science is making rapid progress nowadays. However, the second reason is quite controversial, and for that very reason, rather revealing to hint at Hilbert's original point of view. For, here Hilbert emphasizes that we do have now the "axiomatic method".

---

[7]For a detailed study of the examples Hilbert used for deepening the foundations, see Stöltzner (2002, 2003), for he classifies the examples of deepening the foundations in individual sciences suggested by Hilbert (1918) into eight different types. See also Park (2008, pp. 17–23).

But yet a second circumstances helps us today towards a solution of that old philosophical problem. Not only have the technique of experimentation and the art of erecting theoretical edifices in physics attained new heights, but their counterpart, the science of logic, has also made fundamental progress. Today there is a general method for the theoretical treatment of questions in the natural sciences, which in every case facilitates the precise formulation of the problem and helps prepare its solution—namely, the axiomatic method. (Hilbert 1930, p. 1158, [6])

To be sure, Hilbert did not explicitly mention his own achievements in the area of the axiomatic method. However, in view of the fact that here Hilbert did not mention "Hilbert problems", "Hilbert program", "proof theory", or "metamathe-matics", which are at least equally famous as the trademarks of Hilbert as the "axiomatic method", it is truly significant that Hilbert pinned down the "axiomatic method" as the one achievement he himself was so proud of.

3   The Limitation of Hilbert's Axiomatic Method from Bernays' Point of View

3.1   The Development of Axiomatic Method

3.1.1   The Early Middle Bernays

Ironically, it seems that the later Bernays did not value the axiomatic method highly, as Hilbert did. This interesting point is indeed an idea governing the present paper. For such a difference has all the potentials to become a crucial clue for the almost impossible project of differentiating Bernays' thought from that of Hilbert.[8] Bernays (1922a) is an important paper, in which Bernays evaluated the achieve-ment of Hilbert in his early and middle periods. Since he became an assistant of Hilbert in 1917, we may understand Bernays's early impression of Hilbert in this paper. Bernays started his discussion by briefly scheming how the relationship between mathematics and philosophy had changed in the 18th and the 19th cen-turies since the time of modern scientific revolution. According to Bernays, mathematics imposed strong influence on philosophy until Kant's time. Then, such an influence suddenly disappeared making mathematics and philosophy alienated since the beginning of the 19th century under the spirit of the age of enlightenment. However, in contra distinction to the mainstream philosophy which distanced itself from mathematics, interestingly philosophical orientation of mathematicians evolved continuously. Bernays found the reason for such a development mostly in the fact that mathematics in the 19th century made an overwhelmingly rapid pro-gress. Here, he pinpointed especially the expansion of the research areas, the enlargement? Of the entire structure of research?, and the progress in the method.

---

[8]For example, we have a report of very suggestive episode in Specker (1979, p. 383).

Of course, Bernays was careful enough to mention ground breaking events in the histories of geometry, arithmetic, and algebra. For example, he pointed out that in geometry the spatial intuition became no longer significant epistemologically as the foundation, the concept of number was generalized by the invention of set theory, or the development of algebra came to cover not only numbers and quantity but also the formal systems of mathematics under its purview. In a word, according to Bernays, mathematicians finally won the autonomy of mathematics:

> The bounds the previous philosophical view, and even the Kantian philosophy, had marked out for mathematics were burst. Mathematics no longer allowed philosophy to prescribe the method and the bounds of its research; rather it took the discussion of its methodological problems into its own hands. (Bernays 1922a, p. 190)[9]

Bernays intended to estimate to what extent Hilbert contributed to such a process against that background, and it is not far cry to say that his discussion is almost exclusively devoted to the axiomatic method. Also, here Bernays seems to adopt almost everything Hilbert claimed about the axiomatic method in Hilbert (1918). First, Bernays started his discussion from what Hilbert (1899) achieved. According to him, by its mathematical content, this book not only exerted strong impetus to the subsequent development of mathematics, but also provided us with a methodological turning point to axiomatics. In order to explain this, Bernays densely presented the history of geometry within a few paragraphs. Before everything else, he emphasized that until Kant axioms had been counted as possessing the epistemological character of self-evidence in the axiomatic systems.

> And still Kant held the view that the success and the fruitfulness of the axiomatic method in geometry and mechanics essentially rested on the fact that in these sciences one could proceed from a priori knowledge (the axioms of pure intuition and the principles of pure understanding [Verstand]). (Bernays 1922a, p. 191)

However, the requirement that each of axioms should express a priori knowledge was soon given up, for, as physics developed, empirical propositions and mere hypotheses were selected as axioms of physics. Furthermore, Bernays pointed out that due to the non-Euclidean geometry and the arguments of Helmholtz the view that geometry is nothing but an empirical science increasingly prevailed. According to Bernays, however, even giving up the apriority did not bring up the essential change in the conventional views of the axiomatic method. He thought that a more powerful change was caused by the systematic development of geometry.

Mathematical abstraction guides us to the construction of the comprehensive systems, in which Euclidean geometry can be concretized, far beyond the domains of spatial intuition, thereby the way toward an entirely new way of mathematical speculation that enabled us to consider geometrical axioms from higher point of view. Then, Bernays pointed out that it becomes immediately evident that such

---

[9]The interpretation that counts securing the autonomy of mathematics as the central motivation for Hilbert's thought has been recently emphasized by Franks (2009). It seems important that Bernays mentioned this point.

considerations are irrelevant with the epistemological character of the axioms, and, as is already apparent in Klein's Erlangen programme, the necessity to sharply distinguish between the epistemological problems and the mathematical problems looms large. Only after having explained some such backgrounds, Bernays was able to argue for the significance of Hilbert (1899) as follows:

> The important thing, then, about Hilbert's "Foundations of Geometry" was that here, from the beginning and for the first time, in the laying down of the axiom system, the separation of the mathematical and logical [spheres] from the spatial-intuitive [sphere], and with it from the epistemological foundation of geometry, was completely carried out and expressed with complete rigor". (Bernays 1922a, p. 192)

Of course, Bernays conceded that there always had been the need that in the proofs we should appeal only to the formulated axioms to secure the rigorous foundations of geometry, and never in any fashion to the spatial intuitions. However, Bernays claimed that Hilbert's axiomatics went a step further than eliminating the spatial intuition. For, "[r]eliance on spatial representation is completely avoided here, not only in the proofs but also in the axioms and the concepts" (Bernays 1922a, p. 192). Then, he introduced the so-called the Hilbertian implicit definition into the discussion:

> According to this conception, the axioms are in no way judgments that can be said to be true or false[10]; they have a sense only in the context of the whole axiom system. And even the axiom system as a whole does not constitute the statement of a truth; rather, the logical structure of axiomatic geometry in Hilbert's sense—analogously to that of abstract group theory—is a pure hypothetical one. ... Thus the axiom system itself does not express something factual; rather, it presents only a possible form of a system of connections that must be investigated mathematically according to its internal [innere] properties. (Bernays 1922a, p. 192)[11]

The usual tactic to highlight the contribution of the early and the middle Hilbert to the axiomatic method (as we saw above, that is actually that of Hilbert himself) is to contrast the genetic and the axiomatic methods. Bernays (1922a) was not the

---

[10]Hallett also emphasizes the following points. Contrary to the standard view about the axiomatic system before Hilbert, axioms are no longer true in Hilbert. Furthermore, they are not even judgments that can be either true or false. Even the entire axiom system does not express the truth (Hallett 1995, p. 137). Hallett thinks that such a disagreement over whether axioms can have truth values ultimately originates from the disagreement over the status of axioms. In case we assume that axioms are true, since they must be truths about the primitive concepts that appear in the axioms, a universal logic that can determine the primitive concepts would be pursued. On the other hand, if we reject the assumption that axioms are true, there would be no need for invoking primitive concepts before developing a theory (Hallett 1995, p. 141).

[11]I discuss the problem of implicit definition in Park (2008). See also Chun (2008), Choi (2007). It is Schlick who made this problem famous, and Friedman discusses it in several writings (see, especially Friedman 1992, 1999). However, interestingly enough, Majer underestimates this problem as subsidiary (Majer 2001, 2002). See also Park (2012). We may continue our interest by turning to Frege/Hilbert controversy. See Choi (2009). For the position of Bernays himself, we need to read Bernays (1942). Usually, based on this review, we say that Frege/Hilbert controversy has not been concluded once and for all.

exception to the rule. (see Weyl 1985 as an another example) Elsewhere, I called the problem of whether to emphasize the continuity or the discontinuity of the tradition before Hilbert and the Hilbert's achievement" the Hilbertian dilemma".[12] If we grasp the first horn, as Majer,[13] we might have to give up the clue to understand the rapid progress of science in the late 19th century, i.e., the so-called the second scientific revolution, and especially the utter novelty of modern mathematics. On the other hand, if we grasp the second horn, as Friedman does, we need to explicate rather thoroughly the mysteries involved in implicit definitions. Bernays (1922a) evidently grasped the second horn of the dilemma.

3.1.2 The Later Bernays

Interestingly, the later Bernays distinguished more minutely the stages of development of the axiomatic method by incorporating the results of the collaboration of the later Hilbert and the middle Bernays. For example, in Bernays (1967), he trichotomize the development of the axiomatic method, i.e., (1) material (or pertinent) axiomatics, (2) definitory (or descriptive) axiomatics, and (3) formalized axiomatic. Bernays (1967), as the title "Scope and Limits of Axiomatics" itself indicates implicitly, unlike the early middle Bernays that merely praised the novelty and the importance of Hilbert's axiomatic method, included some deep reflection on the axiomatic method itself based on the experiments and the accumulated knowledge in almost the half century, thereby showing rather a radical change in stance. The big picture is something like this. The two stages, i.e., material axiomatic and definitory axiomatic, seem to correspond roughly with the stage of genetic method and that of the axiomatic method. If so, the formalized axiomatic would go well with Hilbert program, metamathematics, and formalism, which were achieved through the collaboration of the later Hilbert and the middle Bernays. Let us see briefly what Bernays had to say about each stage.

According to Bernays, in order for "collecting and presenting material" to have scientific value, "suitable conceptions, classifications, generalizations, and assumptions" are needed. He understood these in terms of "theoretical aspect", "directive ideas", "conceptuality". And he claimed that the fruitfulness of the axiomatic method goes with the cases in which "the question is one of putting results together, or of evaluating heuristically a problem situation, or also of subjecting assumptions to empirical tests, or even of investigating their inner consistency" (Bernays 1967, p. 188). In order to exemplify these points, he referred to the cases of Mach, Clausius, and Newton. And he added that we can find similar

---

[12]Park (2012) intends to question whether Friedman's discussion of implicit definition was fair to Hilbert's axiomatic method. See Friedman (1992, 1994, 1999, 2001, 2002, 2008, 2011).

[13]For example, Majer claims that the widespread view that Hilbert separated himself from Euclid's axiomatic approach, and aimed at an entirely new axiomatics must be wrong (Majer 2006, p. 158).

instances in biology and theoretical economics. Then, he presented the concept of material axiomatic as follows:

> All the axiomatic systems considered so far have the common trait that they sharpen the statement of a body of assumptions, either for a whole theory or for a problem situation, and of being embedded in the conceptuality of the theory in question; eventually, then, this conceptuality might be somewhat sharpened by it. We might call them "material" or "pertinent" axiomatic. (Bernays 1967, p. 189)

On the other hand, "abstract disciplines developing the consequences of a structural concept described by axioms" belong to what Bernays called "definitory or descriptive axiomatic". (Ibid., p. 190) The methods used in the deepening of foundations gave rise to a kind of axiomatic different from pertinent axiomatic and which one might call definitory or descriptive axiomatic, to which belong abstract disciplines developing the consequences of a structural concept described by axioms. Examples of such concepts are those of group, lattice, and field (Bernays 1967, p. 190). What is interesting is that Bernays pinpointed "multiplicity of equivalent characterization" as "a remarkable circumstance in descriptive axiomatic". For example, according to him, lattice theory can be axiomatized in terms of the concept of equality or that of relation "sub". Thus, the reason why Bernays focused on multiplicity of equivalent characterization can be found in his conclusion like this:

> Hence axiomatization contains elements of arbitrariness. The newer axiom systems for geometry, in particular those of Hilbert and of Veblen, are also descriptive, in contrast with the axiom system of Pasch, which is intended as a pertinent. (Bernays 1967, p. 190)

Interestingly, Bernays characterized the appearance of the formalized axiomatics, which is the final stage in the development of the axiomatic method, simply as transition from axiomatization to formalization under the influence of symbolic logic. Then, he summarized the characteristics of the formalized axiomatic as follows:

> (1) the possible forms of statements in a theory are delimited a priori, and
> (2) the logical inferences are subjected to explicit rules. (Bernays 1967, p. 191)

On the other hand, Bernays discussed the limitations of the formalized axiomatic in much more detailed fashion. According to him, we can formalize the previous proofs in various areas of mathematics, such as the theory of number, analysis, set theory, and geometry, in the way of formalized axiomatic. However, he claimed: "Nevertheless formalized axiomatic cannot fully replace descriptive axiomatic".

3.2 The Problem of Uniformity of the Axiomatic Method

In the previous section, I tried to reveal the point that, though quite subtle, the early middle Bernays and the later Bernays betrayed some significant differences regarding the axiomatic method. By doing so, I intended to show that Bernays' case can be contrasted in interesting ways with the case of Hilbert, who, regardless of the early, middle, and later period, sustained the firm belief in the axiomatic method, and showed strong affection and pride toward what he contributed to its development. However, the most salient issue, in which the difference between Hilbert and Bernays looms large, must be the problem of the uniformity of the axiomatic method. For, while Hilbert never had any slight doubt on the uniformity of the axiomatic method throughout his career, at least the later Bernays was extremely skeptical about it. So, it becomes mandatory to look into what changes were there in Bernays' position on the problem of the uniformity of the axiomatic method from the early middle period through the late middle period to the later period.

3.2.1 The Early Middle Bernays

Bernays (1922a), after having pointed out that in geometry purely mathematical part, i.e., mathematical structures, focuses on logical relations disregarding the problem of factual truth, discusses the problem of the uniformity of the axiomatic method:

> This sort of interpretation of the axiomatic method presented in Hilbert's "Foundations of Geometry" offered the particular advantage of not being restricted to geometry but of being directly applicable to other disciplines. From the beginning, Hilbert envisaged the point of view of the uniformity of the axiomatic method in its application to the most diverse domains, and guided by this viewpoint, he tried to bring this method to bear as widely as possible. In particular, he succeeded in grounding axiomatically the kinetic theory of gases as well as the elementary theory of radiation in a rigorous way. (Bernays 1922a, p. 193)[14]

Then, Bernays immediately invoked Hilbert (1918), i.e., Hilbert's famous Zürich lecture. For, here Hilbert presented the synopsis of the guiding principles for the study of axiomatics and the overview of the results gotten so far. In a word, Bernays believes that this lecture characterized the axiomatic method as "a general procedure for scientific thinking". Probably, in order to avoid possible misunderstanding, Bernays once more explicitly mentioned the novelties in Hilbert's axiomatics, even by using an interesting expression like "the old axiomatics". Unlike the old axiomatics, Hilbert's axiomatics disregards "the epistemic character of the axioms", and considers "the whole framework of concepts ... only in its internal structure". As a consequence, the theory becomes "the object of a purely mathematical investigation, exactly what is called *axiomatic* investigation", and exactly here problems of proving the consistency and independence of the axiomatic systems are raised. Then, finally Bernays discussed "deepening the foundations", which is one of the central issues in Hilbert (1918):

> In addition, there remains the task of investigating the possibilities of a "deepening of the foundations," that is, examining whether the given axioms of the theory might not be reduced to propositions of a more fundamental character that would then constitute "a

---

[14]It is interesting that there is Bernays (1922b) among his writings.

deeper layer of axioms" for the framework of concepts under consideration. (Bernays 1922a, p. 193)

Bernays claimed that studies such as these certainly have mathematical characters, and can be applied to any domain of knowledge where theoretical treatment is appropriate. They have utmost value in the clarification and the systematic overview of knowledge. This means that the early middle Bernays was happily adopting the ideal of "deepening the foundations" of the middle Hilbert. This ideal symbolizes Hilbert's ambitious scheme for the status and the roles of mathematics in the kingdom of sciences. Since it is the axiomatic method that enables us to realize this scheme, insofar as he adopted this ideal, Bernays also shared Hilbert's affection and belief in the axiomatic method. In fact, Bernays claimed: "Thus mathematical thought gains a universal significance for scientific cognition [Erkennen] through the idea of axiomatics". Further, he even claimed:

> By means of this comprehensive development [Ausgestaltung] of axiomatic thought, a sufficiently wide context was indeed obtained for the mathematical formulation of problems, and the epistemological fruitfulness of mathematics was made clear. (Bernays 1922a, p. 194)

3.2.2   The Later Bernays

Now let us take a look at what Bernays' views on the uniformity of the axiomatic method based on Bernays (1967). In view of the stages of development in the axiomatic method discussed above, and the expressed views of Bernays in Bernays (1922a), the ideal of "deepening the foundations" seems to be pursued by presupposing the axiomatic method. In other words, "deepening the foundations" seems to take place only after the material (pertinent) axiomatics and definitory (descriptive) axiomatics appeared. Surprisingly, however, the later Bernays understood the matter as if the ideal of "deepening the foundations" is in action in the transitional stage between material axiomatics and definitory axiomatics. First, Bernays pointed out that the important events in theoretical sciences, especially in physics, possess one axiomatic character. What he was invoking here are the cases in which one discipline is specified into another discipline or two separate disciplines are assimilated into one. He called these cases "reduction".

According to him, we cannot say that simply axiomatics is involved, for these reductions should be motivated and justified by in part experimental results. Nevertheless, he thought that some such reductions could be considered from the perspective of the axiomatic method, which is of the same sort as what Hilbert called "deepening the foundations". And, he claimed that the methods used in deepening the foundations brought up the axiomatics, which is different from material axiomatics, and we may call it definitory or descriptive axiomatics, to which belong abstract scientific disciplines that develop the consequences of the structural concepts described by the axioms (Bernays 1967, p. 190).

Now we need to remember that at the beginning of Bernays (1967) he remarked rather shockingly that "there often seems to be not so much a need for recommending axiomatic as to warn against an overestimation of it". Who were those who deserved to be warned by their overestimation of axiomatic? In fact, we don't have to speculate about it, for Bernays himself explicitly mentioned those he had in mind:

> In fact, there are today mathematicians for whom science begins only with axiomatic, and there are also mathematically minded philosophers, especially in Carnap's school, who regard axiomatization as belonging to the construction of scientific languages. (Bernays 1967, p. 188)

It must be the homework for the future historians to identify the mathematicians and scientists Bernays had in mind. Before everything else, however, we should probe the question as to why Bernays attacked Carnap so vehemently.

4   Bernays' Criticism of Carnap

In a recent article, which will be the point of departure for any study on Bernays' thought, Charles Parsons critically examined Bernays' philosophical thought after 1945 (Parsons 2006). He focused on anti-foundationalism, the denial of a priori knowledge, and structuralism, as the typical characteristics of Bernays. His overall assessment is somewhat negative, for he expressed disappointment about meager achievement of Bernays, contrary to the expectation from the sheer quantity of the writings (Parsons 2006, p. 148).[15] From my point of view, the issues Parsons focused on are not only philosophically important but also timely in that they could have far-reaching implications to today's advanced studies. However, the problem is that, if we focus on them, the difference between Hilbert and Bernays may not be revealed clearly.[16] And, I believe that such a wrong impression was resulting by Parson's failure to discuss the problem of the uniformity of the axiomatic method, which is indispensable for understanding Bernays' originality. This point of view

---

[15]Parsons also complains about Bernays' essay style of writing, which lacks systematic approach. Further, he thinks that Bernays' discretion in discussing different positions and his loyalty to those he had close friendship tend to make his position as close to syncretism (Parsons 2006, p. 148).

[16]Above all, Parsons seems to discuss Bernays' anti-foundationalist position by assuming that Hilbert's studies in foundations of mathematics at least indirectly makes him a foundationalist. However, the recent study of Franks shows persuasively that there are notably anti-foundationalist elements in Hilbert himself (see Franks 2009, pp. 32–40). The problem of the critique of a priori knowledge is an old problem that provoked in-depth discussion for a long time within the Hilbert school. It shows much more complex modes than one might think. However, if we have in mind Nelson's criticism of Hilbert, and Bernays' rebuttal, there is no doubt that there are almost insurmountable difficulties in contrasting Hilbert and Bernays through this problem (see Bernays 1928). The case of structuralism no doubt shows affinities not only with Bernays but also with Hilbert. In fact, many mathematical structuralists of our time trace their views back to Hilbert. Thus, here again it is difficult to highlight the differences between Hilbert and Bernays.

provides us with an understandable solution to the problem as to "why did Bernays identified Carnap as the typical one who overestimate the axiomatic method and counting axiomatization as belonging to the construction of scientific language?" This is so, because, above all, both the fact that Carnap had a faith in the uniformity of the axiomatic method and another fact that Carnap counted axiomatization as belonging to the construction of scientific language agree with the position of the later Carnap after *Logical Syntax*. However, immediately the following question will be raised.

[Q]: How could Bernays, who was a long-time collaborator of Hilbert, attack Carnap for the reason that he adopts the trademark of Hilbert, i.e., the uniformity of the method of the axiomatic method?

This question may be understood as the conjunction of the following small questions: (1) Isn't Carnap a representative person who succeeded and continued the tradition of logicism? (2) As we saw above, isn't the uniformity of the axiomatic method one of Hilbert's beliefs? (3) Wasn't Bernays, as a collaborator of Hilbert, a champion of formalism? As for the question (1), Beth (1963) and Carnap (1963), both published in Schillp (1963), may give us a rough answer. Beth viewed Carnap's Logical Syntax as an attempt to reconcile logicism and formalism. However, he attacked Carnap as follows: "Rather than a fusion of logic and formalism or an incorporation of formalism into logicism, *Logical Syntax* appears to advocate a surrender of logicism to formalism" (Beth 1963, pp. 475–476). In response to this, Carnap, after having conceded that Beth's comments were extremely interesting. And generally he agreed with them, presented the following suggestive answer:

> But it seems to me advisable to make a clearer distinction between formalism and the formalist method. The *formalist method*, or in my terminology the syntactical method, consists in describing a language L together with its rules of deduction by reference only to signs and the order of their occurrence in expressions, thus without any reference to meaning. The application of the formalist method in the construction of a language L does not in itself exclude adding an interpretation for L, but if we do so, this interpretation does not enter into the syntactical rules for L. *Formalism*, in the sense of the conception about the nature of mathematics represented by Hilbert and his followers, consists of both the proposal to apply the formalist method and, more essentially, the *thesis of formalism*, that this is the only possible way of constructing an adequate system of mathematics, since it is impossible to give an interpretation for (classical) mathematics. In this assertion the *thesis of logicism*, that all terms of mathematics can be interpreted in terms of logic, is rejected. I accepted the formalist method and developed it in a wider domain, but I did not accept the thesis of formalism and instead maintained that of logicism. (Carnap 1963, p. 928)

Bohnert (1975) also reported the same opinion as Carnap's statement just quoted.

On the other hand, as far as the questions (2) and (3) are concerned, we may resolve the tension by resorting to the discussion above. The later Bernays, unlike his previous self, became very skeptical about the uniformity of the axiomatic method. And he estimated the place of Hilbert within the history of the axiomatic method with much more systematic and rigorous views about the stages in the development of the axiomatic method. If so, have we secured a satisfactory answer to the question ([Q]: *How could Bernays, who was a long-time collaborator of Hilbert, attack Carnap for the reason that he adopts the trademark of Hilbert, i.e., the uniformity of the method of the axiomatic method*?)? Perhaps, not. Thus, we need to look into Bernays (1961) to check whether we can gather more clues from it. For example, Bernays wrote:

> The general tendency of the Logical Syntax can be said to be an extension of the approach of Hilbert's proof theory. For Hilbert the method of formalization is applied only to mathematics. However, in his lecture "Axiomatic Thought" Hilbert also said: "Everything at all that can be the object of scientific thinking falls under the axiomatic method, and thereby indirectly under mathematics, when it becomes mature enough to form into theory". Carnap goes a step further in this direction in the Logical Syntax, by considering science as a whole as an axiomatic deductive system which becomes a mathematical object through formalization: the syntax of the language of science is metamathematics that is directed towards this object. (Bernays 1961, 186; I am borrowing the unpublished English translation in Bernays Project, p. 2)

Though it appears clear enough, this paragraph contains many troubling elements. Here Bernays seems to criticize Hilbert for considering the axiomatic method as peculiar to pure mathematics. On the other hand, Bernays seems to criticize Carnap as unjustifiably expanding the axiomatic method to the entire science. However, such an interpretation may be refuted by the quoted paragraph itself. For, there seems to be no difference at all between the position of Hilbert (1905) and that of Carnap. Of course, one may avid the criticism by pointing out that Bernays did not claim that "For Hilbert the method of the axiomatic method is applied only to mathematics" but merely claimed "For Hilbert the method of formalization is applied only to mathematics". However, in such a case, he would be smuggling the much clearer content he suggested later in his discussion of formalized axiomatic as the final stage in the development of the axiomatic method. (see Bernays 1967) If so, is it impossible to interpret this quotation as targeting both Hilbert and Carnap, though in appearance it does not aim at Hilbert as a target?

5   Concluding Remarks

By highlighting the point that Hilbert was so proud of his own axiomatic method on textual evidence, in this article, I tried to understand Bernays' views about the place of Hilbert's axiomatic method in the history of the axiomatic method. Taking clues from the point that the middle Bernays and the later Bernays took very different positions about this problem, I showed the views of the middle Bernays with that of the later Bernays. And, as a consequence, I pointed out that the difference between the views of Hilbert and Bernays can be traced back to the problem of the uniformity of the axiomatic method. On the same line of thought, I presented some preliminary discussions of the later Bernays' criticism of Carnap, by interpreting

Carnap's project[17] in philosophy of science after the mid 1950s as an endeavor to succeed Hilbert's belief in the uniformity of the axiomatic method. I failed in this article to inquire by what process Bernays' views on the axiomatic method went through the changes discussed above. Nor did I uncover how the several salient characteristics of Bernays' later philosophy are related to the differences between Hilbert's and Bernays' views on the axiomatic method. Unfortunately, I have to leave these questions for other occasions.

In lieu of the conclusion, I would like to hint at some more speculative ideas. Did Bernays view Carnap's work as a distortion of Hilbert's position? Wasn't he himself distorting Hilbert's position? If so, was it an intentional distortion or a mere misunderstanding? If not, did his critical reflection on Carnap provide him with an opportunity to sharply distinguish between Hilbert's and his own positions? Further, speculations like this is naturally led to express suspicion about some common senses in the history of logic and mathematics. To what extent the so-called Hilbert program, which was born by the collaboration of Hilbert and Bernays, represent Hilbert's position? To what extent is Hilbert's position prior to 1920, especially Hilbert's belief in deepening the foundations, and the axiomatic method connected to Hilbert program? Isn't Hilbert program in fact Bernays program? If so, shouldn't we view Carnap's work as an attempt to continue Hilbert's view on the axiomatic method (without adopting Hilbert program or Bernays program)? Wasn't Bernays' criticism of Carnap a repulsion to Carnap's some such attempt? Who was the true successor of Hilbert, if not Weyl, von Neumann, Gödel, or Bernays? If the axiomatic method was so important to Hilbert, shouldn't its spirit be continued at least by some adopted children?

References

Bernays, P. (1913). Über die Bedenklichkeiten der neueren Relativitätstheorie. *Abhandlungen der Fries'schen Schule, IV*(3), 459–482.
Bernays, P. (1922a). Die BedeutungHilbertsfür die Philosophie der Mathematik. *Die Naturwissenschaften 10*, 93–99 (English translation as Hilbert's Significance for the Philosophy of Mathematics 189–197, in Mancosu (1998a)).
Bernays, P. (1922b). Zur mathematischen Grundlegung der kinetischen Gastheorie. *Mathematische Annalen, 85*, 242–255.
Bernays, P. (1928). Uber Nelsons Stellungnahme in der Philosophie der Mathematik. *Die Naturwissenschaften, 16*, 142–45 (On Nelson's Position in the Philosophy of Mathematics, Bernays Project: Text No. 7).

---

[17]The prime examples of this project are Carnap (1956, 1958/1975, 1959/2000), and they were discussed by Carnap (1966/1974) in sufficiently detailed fashion. Carnap (1961) must be extremely important for understanding the relationship between Hilbert and Carnap. Friedman (1992) views Carnap's later project in philosophy of science was already budding in Carnap (1934, 1939). See Friedman (2008, 2011) for how later Carnap expanded the idea of Ramsey sentence by the impetus from Hempel.

Bernays, P. (1942). Review of Max Steck, 'Einunbekannter Brief von Gottlob Gregeüber Hilbertserste Vorlesungüber die Grundlagen der Geometrie. *Journal of Symbolic Logic, 7*(2), 92–93.

Bernays, P. (1961). Zur Rolloe der Sprache in erkenntnistheoretischer Hisicht. *Synthese, 13,* 185–200 (On the Role of Language from an Epistemological Point of View. Bernays Project: Text No. 25).

Bernays, P. (1967). Scope and limits of axiomatics. In M. Bunge (Ed.), *Lecture at the Delaware seminar in the foundations of physics* (pp. 188–191). Berlin, Heidelberg, New York: Springer.

Beth, E. (1963). Carnap's views on the advantages of constructed systems over natural languages in analytic philosophy. In Schillp (Ed.) (pp. 469–502).

Bohnert, H. G. (1975). Carnap's logicism. In Hintikka (Ed.) (pp. 183–216).

Carnap, R. (1934/1937). *The logical syntax of language.* London: Kegan Paul.

Carnap, R. (1956). The methodological character of theoretical concepts. In H. Feigl & M. Scriven (Eds.), *Minnesota studies in the philosophy of science, vol. I: The foundations of science and the concepts of psychology and psychoanalysis* (pp. 38–76). Minneapolis: University of Minnesota Press.

Carnap, R. (1958/1975). Observational language and theoretical language (pp. 75–85). In J. Hintikka (Ed.) (1975) (Translated by H. G. Bohnert from *Dialectica, 12*(1958), pp. 236–248).

Carnap, R. (1959/2000). *Theoretical concepts in science* in Psillos (2000) (pp. 158–172).

Carnap, R. (1961). On the use of Hilbert's e-operator in scientific theories. In Y. Bar-Hillel, et al. (Eds.), *Essays on the foundations of mathematics* (pp. 156–164). Jerusalem: The Magnus Press.

Carnap, R. (1963). *Replies and systematic expositions,* in Schillp (1963) (pp. 859–1013).

Carnap, R. (1966/1974). *An introduction to the philosophy of science.* New York: Basic Books.

Choi, W. (2007). Hume's principle and implicit definition. *Korean Journal of Logic, 102,* 23–46. (in Korean).

Choi, W. (2009). Frege-Hilbert controversy about existence and consistency. *Philosophy, 99,* 127–148. (in Korean).

Chu, Y. S. (2008). Hilbert and Carnap in the theory of space. *Korean Journal for the Philosophy of Science, 11,* 1–134. (in Korean).

Chun, Y. S. (2008). Hilbert and Carnap on the theory of space, *Korean Journal for the Philosophy of Science,, 11,* 1–35.

Corry, L. (2004). *David Hilbert and the axiomatization of physics (1898–1918).* Dordrecht: Kluwer.

Feferman, S. (2008). Lieber Herr Bernays!, Lieber Herr Gödel! Gödel on finitism, constructivity, and Hilbert's program. *Dialectica, 62*(2), 179–203.

Franks, C. (2009). *The autonomy of mathematical knowledge: Hilbert's program revisited.* Cambridge: Cambridge University Press.

Friedman, M. (1992). Carnap and a priori truth. In D. Bell & W. Vossenkuhl (Eds.) (pp. 47–60).

Friedman, M. (1994). Geometry, convention, and the relativized A priori. In W. Salmon & G. Wolters (Eds.) (pp. 21–34) (Reprinted in Friedman (1999), Chap. 3, pp. 59–70).

Friedman, M. (1999). *Reconsidering logical positivism.* Cambridge: Cambridge University Press.

Friedman, M. (2001). *Dynamics of reason.* Stanford, California: CSLI Publications.

Friedman, M. (2002). Geometry as a branch of physics: Background and context for Einstein's 'geometry and experience'. In D. B. Malament (Eds.) (pp. 193–229).

Friedman, M. (2008). *Wissenschaftslogik*: The role of logic in the philosophy of science. *Synthese, 164,* 385–400.

Friedman, M. (2011). Carnap on theoretical terms: Structuralism without metaphysics. *Synthese, 180,* 249–263.

Hallett, M. (1995). Hilbert and logic. In M. Marion & R. S. Cohen (Eds.), *Quebec studies in the philosophy of science,* Part I (pp. 135–187). Dordrecht: Kluwer.

Hilbert, D. (1899/1990). *Grundlagen der Geometrie,* Teubner, Leibzig (English Translation of the Tenth German edition, *Foundations of Geometry,* LaSalle: Open Court).

Hilbert, D. (1900). Ueber den Zahlbegriff. *Jahresbericht der Deutschen Mathematiker-Vereinigung, 8*, 180–184. (English translation in Ewald (1996), pp. 1089–1096).

Hilbert, D. (1905). Ueber die Grundlagen der Logik und der Arithmetik. In A. Krazer, (Ed.) *Verhandlungen des dritten Internationalen Mathematiker-Kongresses in Heidelberg vom 8. bis 13. August 1904* (pp. 174–85). Leipzig: Teubner (English translation in van Heijenoort (1967), pp. 129–38).

Hilbert, D. (1918). AxiomatischesDenken. *MathematischeAnnalen, 78*, 405–415. Lecture given at the Swiss Society of Mathematicians, 11 September 1917 (English translation in Ewald (1996), pp. 1105–1115).

Hilbert, D. (1930). Naturerkennen und Logik. *Die Naturwissenschaften, 18*, 959–963 (English translation in Ewald (1996), pp. 1157–1165).

Hilbert, D., & Bernays, P. (1934). *Grundlagen der Mathematik* (Vol. 1). Berlin: Springer.

Hilbert, D., & Bernays, P. (1939). *Grundlagen der Mathematik* (Vol. 2). Berlin: Springer.

Hintikka, J. (Ed.). (1975). *Rudolf Carnap, logical empiricist*. Dordrecht: Reidel.

Majer, U. (2001). *The axiomatic method and the foundations of science: Historical roots of mathematical physics in Göttingen (1900–1930)*. In M. Redei & M. Stöltzner (Eds.) (pp. 11–33).

Majer, U. (2002). Hilbert's program to axiomatize physics (in analogy to geometry) and its impact on Schlick, Carnap and other members of the Vienna Circle. In Heidelberger and F. Stadler (Eds.) (pp. 213–224).

Majer, U. (2006). *Hilbert's axiomatic approach to the foundations of science—a failed research program?* In V. F. Hendricks (2006) (pp. 155–184).

Mancosu, P. (Ed.). (1998a). *From Brouwer to Hilbert. The debate on the foundations of mathematics in the 1920s*. Oxford: Oxford University Press.

Mancosu, P. (1998b). *Hilbert and Bernays on metamathematics*. In P. Mancosu (Ed.) (1998a) (pp. 149–188).

Park, W. (2008). Zermelo and the axiomatic method. *Korean Journal of Logic, 11–2*, 1–57.

Park, W. (2011). Bernays and the axiomatic method. *Korean Journal of Logic, 14*(2), 1–37. (in Korean).

Park, W. (2012). Friedman on implicit definition: In search of the Hilbertian heritage in philosophy of science. *Erkenntnis, 76*(3), 427–442.

Parsons, C. (2006), Paul Bernays' Later philosophy of mathematics. *Logic colloquium 2005*. In C. Dimitracopoulos et al. (Eds.), Lecture Notes in Logic 28. Associations for Symbolic Logic. Cambridge: Cambridge University Press.

Reid, C. (1996). *Hilbert*. Dordrecht: Springer.

Schilpp, P. A. (Ed.). (1963). *The philosophy of Rudolf Carnap*. La Salle: Open Court.

Specker, E. (1979). Paul Bernays, *Logic colloquium 78*. In M. Boffa et al. (Eds.). Amsterdam: North-Holland Publishing Company.

Stöltzner, M. (2002). How metaphysical is deepening the foundations? Hahn and Frank on Hilbert's axiomatic method. In M. Heidelberger & F. Stadler (Eds.) (pp. 245–262).

Stöltzner, M. (2003). The principle of least action as the logical empiricist's Shibboleth. *Studies in History and Philosophy of Modern Physics, 34*, 285–318.

Weyl, H. (1985). Axiomatic versus constructive procedures in mathematics. In T. Tonietti (Ed.), *The mathematical intelligencer, 7*(4), 10–17 and 38.

Zach, R. (1999). Completeness before post: Bernays, Hilbert, and the development of propositional logic. *The Bulletin of Symbolic Logic, 5*(3), 331–366.

Part III Goedel and Tarski

Part III Gödel and Tarski

Chapter 8
Patterson on Tarski's Definition of Logical Consequence

Abstract We still do not know against what historical/philosophical background and motivation Tarski's definition of logical consequence was introduced, even if it has had such a strong influence. In view of the centrality of the notion of logical consequence in logic and philosophy of logic, it is rather shocking. There must be various intertwined reasons to blame for this uncomfortable situation. There has been remarkable progress achieved recently on the history of analytic philosophy and modern logic. In view of the recent developments of the controversies involved, however, we will have to wait years to resolve all these uneasiness. In this gloomy situation, Douglas Patterson's recent study of Tarski's philosophy of language and logic seems to have the potential to turn out to be a ground breaking achievement (Patterson 2012). This chapter aims at uncovering the state-of-the-art and fathoming the future directions of the research in this problem area by examining critically some unclear components of Patterson's study.

Keywords Alfred Tarski · Douglas Patterson · Intuitionistic formalism
Logical consequence · Rudolf Carnap

1 Introduction

We are still lacking enough understanding of the historical and philosophical background and motivation of Tarski's definition of the concept of logical consequence. This is rather shocking, since the concept of logical consequence is central in logic and philosophy of logic. Several factors may have played in complex ways for resulting such an unsatisfactory situation. Though recent achievements in the history of analytic philosophy and logic are certainly encouraging, the development of the on-going controversies indicates that it will take much time to overcome the unsatisfactory situation. Douglas Patterson's recent book on Tarski's philosophy of language and logic seems potentially a revolutionary achievement against such a

An earlier version was published in Korean as Park (2014).

gloomy background (Patterson 2012). This chapter aims at highlighting some of the most remarkable findings of Patterson's study of Tarski as well as critically examining some dubious aspects in it. Hopefully, we may understand exactly where we are and what to do in our future endeavor.

No one would deny that model theory was largely established by Tarski and his students. Tarski's definition of truth in formalized languages was the foundation stone for building model theory. Also, the definition of logical consequence and the model-theoretic definitions of closely related notions such as logical truth and consistency started from Tarski's 1936 article (Tarski 1983, pp. 409–420).[1] Of course, the definition of logical consequence suggested by Tarski in the 1930s is not exactly the same as our current model-theoretic definition of logical consequence. However, since Tarski's definition of logical consequence suggested in the 1950s seems almost the same as the currently accepted one, and Tarski himself did not comment on why and how he revised his own definition of logical consequence, usually we invoke the Tarskian model-theoretic definition of logical consequence without any clear distinction between these definitions.[2] Thus, it is still an open problem for historians and philosophers of logic to understand what motivated the Tarskian model-theoretic definition of truth to appear, what tortuous paths it had to go through, and what urgent philosophical issues are still left (see Asmus and Restall 2012).

In order to show that we are still lacking enough understanding of the motivation, background and the purpose of Tarski's 1936 article on logical consequence, let me refer to two examples. First, Jané aptly shows not only that Tarski indeed made a sharp distinction between "common concept", "formalized concept", and "proper concept", but also there is a need to clarify the relationships between the first and the third in his 1936 article (Jané 2006). Surprisingly, Jané concludes that the common concept of logical consequence in Tarski is not a certain general concept of consequence but an exact concept of consequence used in axiomatics. Secondly, in their first English translation of Tarski's 1936 article from the Polish original, Stroinska and Hitchcock recently treat Carnap's attempt to define logical consequence as an example of semantic approach. To say the least, such an interpretation is potentially controversial (Stroinska and Hitchcock 2002).[3] Even though Tarski in his 1936 article explicitly mentioned Carnap as the first instance of attempting to define logical consequence, there is no evidence that he counted it as a semantic work. Tarski was referring to Carnap's *Logical Syntax* published in 1934,

---

[1]*"The sentence X follows logically from the sentences of the class K if and only if every model of the class K is also a model of the sentence X"* (Tarski 1983, p. 417). For a brief history of model theory, see Vaught (1974, 1986).

[2]"A sentence Φ is said to be a logical consequence of a set A of sentences if it is satisfied in every realization R in which all sentences of A are satisfied" (Tarski et al. 1971, p. 8). See also Tarski and Vaught (1957).

[3]See the outline of Tarski's paper presented by Stroinska and Hitchcock: "0. Introduction, 1. The syntactic approach ..., 2. The semantic approach, 2.1. Carnap's definition (1934) ..., 3. Logical terms..." (Stroinska and Hitchcock 2002, 158–160).

and at that time Carnap was not yet adopting semantic approach to logical consequence. This can be evidenced by the autobiographical remark of Carnap himself about his initial response to Tarski's intention to present on the definition of logical consequence at 1935 Paris Congress (Carnap 1963, p. 60).

In order to understand the significance of the Tarskian model-theoretic definition of logical consequence, before everything else we need to grasp in what respects it is superior to the previous syntactic definition. Probably, that was the most important issue, when Tarski first suggested it. After all these years, however, the problem itself seems to be forgotten, since the Tarskian approach has been the dominant stream for more than 70 years. Of course, Tarski himself is partially responsible for the fact that we are failing to understand the original intention, motivation, and the historical background of Tarski's definition. Tarski discussed in the footnotes rather meticulously who should be credited for which concrete achievement in relation to the problems around the semantic approach. Contrastingly, he was extremely reticent in giving information about the syntactic approach. After all, who was the true target of Tarski's criticism of syntactic approach? Why was he so silent about this matter?

Admittedly, even without any explicit mention, scholars at that time might have known who or which school was the target of Tarski's criticism of syntactic approach. One might consider the fact that Tarski found from Gödel's incompleteness theorem the ultimate failure of the syntactic approach as indicating indirectly and implicitly that Hilbert and his followers were the targets. Or it could be the reason for Tarski's silence that in early years he himself was one of those who took syntactic approach. As Bach (1997) points out, Tarski, in his 1931 article on truth and 1933 article on ω-consistency, defined consequence relation syntactically. No matter what reason was there, it is unfortunate for the historians of logic that they cannot uncover it. At least, we should be able to correlate some of the remarkable achievements in the history of syntactic approaches with particular authors' particular writings.

We may cite Woleński's series of writing as the typical study of the scientific background of Tarski with all these problems in mind (Woleński 1989, 1995, 1999). For example, we may get the following impression from Woleński (1995). First of all, we should not forget the fact that Tarski was a mathematician rather than a philosopher. It is also clear that, as a mathematical logician, he was concerned much more to establish the nascent mathematical logic as a genuine mathematical discipline than to interact actively with philosophers via philosophy of logic (see also Sinaceur 2009, p. 370). Though he learned and digested a lot from each and every one of logicism, formalism, and intuitionism, he never ascribed his own stance to any one of these schools. Even if the influences of his Polish teachers to him cannot be ignored, it is rather difficult to interpret him as the follower of any one particular scholar among Leśniewski, Kotarbiński, and Lukasiewicz. (See Betti 2008, Patterson 2008, Sundholm 2003) There is no doubt that he was deeply interested in philosophical issues, and well versed in the history of philosophy. However, from his writings, we get the impression that he was anxious to avoid philosophical discussions. To make the matters worse, he had to take entirely different styles of exposition depending on the target readers. As a consequence, sometimes we are bound to doubt whether Tarski was in fact inconsistent.[4]

In order to understand appropriately the historical and philosophical background of Tarski's model-theoretic definition of truth, it is mandatory to look into the details of all the scholars and the schools mentioned above. However, it seems also necessary to avoid the stupidity of losing sight of the whole by minding too much of the small details. If so, important clues for if our research should be the following:

> Clue (1): Tarski, in his 1936 article, emphasized that his own semantic definition is superior to the previous syntactic definition. (Tarski 1983, 2002)
>
> Clue (2): When Tarski presented on the semantic concept of consequence at 1935 World Congress on Scientific Philosophy, many logical positivists, including Neurath vehemently opposed. (Carnap 1963, p. 61)
>
> Clue (2'): Due to the World War II, many European scholars emigrated to U.S.A. Thanks to this, in 1941, there was a monumental event in the history of logic. At Harvard, Russell, Carnap, Tarski, Quine, and some other distinguished logicians such as Goodman had a regular seminar. (Frost-Arnold 2013)
>
> Clue (3): In his early years, Tarski named his own position "Intuitionistic Formalism". (Tarski 1983, p. 62)[5]

I interpret Clue (1) as indicating the necessity to understand Tarski's definition in the light of Hilbertian thought. Even if Tarski learned from each and every one of the three big schools, i.e., logicism, formalism, and intuitionism, largely speaking, his thought shows most affinity with formalism. Also, only against the background of Hilbertian proof theory, we may understand the basic motivation of model theory Tarski was newly establishing. In view of all this, it is rather surprising that the relationship between Tarski's achievements and Hilbertian thought has been rarely studied.[6]

On the other hand, I interpret clue (2) and clue (2') as indicating the necessity of understanding Tarski's definition in the light of logical positivist thought. It is well known that, at Harvard seminar, Quine and Tarski were extremely critical about Carnap's intention to sharpen the distinction between logical truth and factual truth (or the distinction between logical and extra-logical). Furthermore, Tarski's interest in these distinctions and the disagreement of Tarski and Carnap about these can be traced back to earlier decade. Spontaneously, we turn to the heated debates and

---

[4]It is not hard to find confirming instances for such an impression. See Feferman and Feferman (2004). Their report that Tarski tended to be reluctant to express his opinions on philosophical problems is a clear example.

[5]As Woleński points out, "intuitive formalism" may be a better name for avoiding any confusion with Brouwer's intuitionism (Woleński 1995, p. 336). However, I will use "intuitionistic formalism" throughout this paper following Tarski's usage.

[6]This point can be supported strongly by the fact that there is no mention of Hilbert in the index of Feferman and Feferman's biography of Tarski (Feferman and Feferman 2004). As far as I know, Sinaceur is the almost only case that paid due attention to the relationship between the Hilbert school and Tarski before Patterson (2012) (see Sinaceur 2009).

controversies within the logical positivist camp, which has been sometimes called "Logical positivists' dilemma". When logical positivists were torn between the correspondence theory and the coherence theory, Tarski intervened in such a way that Carnap had to change his position significantly. In such a situation, it cannot be a coincidence that Neurath showed more focal interest than anyone else.

Clue (1), Clue (2), and Clue (2') are all containing ample information for us to understand Tarski's model-theoretic definition of logical consequence. So, what is important is to note that the interrelationship among these clues reveals significant information. For example, in-depth understanding the relationship between the Hilbert school and logical positivism could be one possible route to understanding Tarski appropriately. The rise of logical positivism was almost simultaneous with the fall of the Hilbert School. That cannot be a mere historical accident. If we note the series of events that some of the most important logical positivists, such as Carnap, Schlick, and Reichenbach, became famous in some connection with Einstein's marvelous success, that after Einstein/Hilbert priority debate in 1916 as the pinnacle Hilbert became declined, and that to Hilbert who became more dependent on Bernays, that Gödel's incompleteness result is 1931 was a crucial blow, it is evident that we need to study Hilbert together with logical positivism. "Deepening the foundations" is another trademark of Hilbert, which was prominently emphasized by him before he launched the so-called Hilbert program in collaboration with Bernays. It was a project of applying the axiomatic method in mathematics to all the individual sciences. No doubt, this project seems to have remarkable similarities with logical positivists' project of unified science.[7]

Clue (3) must have an undeniable authority as an auto-biographical report of Tarski. However, it seems difficult to find more detailed information except for the fact that Tarski himself explicitly made it clear that he no longer subscribed to Intuitionistic Formalism (Tarski 1983, p. 62, footnote †). As Woleński laments, "[u]nfortunately, Tarski neither explained why he rejected his former view nor enlarged on his later attitude" (Woleński 1995, p. 336).

Now we note the fact that Douglas Patterson's recent book seems to have the potential to explain persuasively and coherently the whole story of what happened by synthesizing all the clues we enumerated above. Above all, the excellence of Patterson is found in that he relatively clearly highlights the meaning and importance of Clue 3, which has been in front of us through all these years. And, that was possible as a result of investigating patiently how Clue 3 is interrelated organically

---

[7]From this point of view, it is good news that there is on-going publication of Hilbert's lecture notes and Bernays' collected work. The reevaluation of logical positivism for the last three decades is also quite encouraging. Even if we confine our interest to Tarski, we may witness similar phenomena such as the publication of Feferman and Feferman's biography of Tarski, Woleński's history of modern Polish philosophy, and some in-depth studies of modern history of logic by Mancosu, Sinaceur, and Patterson. Though it is beyond the scope of this paper, it must be also interesting and promising to understand the role played by Gödel and Tarski against the big picture in which the allied force of logical positivism (in Vienna, Berlin, and Prague) and the Lvov-Warsaw school (in Poland) was attacking the Hilbert school.

with other clues. As far as Clue (1) and Clue (3) are concerned, if Intuitionistic Formalism is a kind of formalism, it seems natural to ask in what respects it is similar to and different from Hilbertian formalism or the formalistic aspects of Carnap. And, Patterson indeed gave serious effort to answer these questions. To that extent, we may safely believe that Patterson examined seriously the connection between Clue (1) and Clue (3). In the case of Clue (2) and Clue (3), again we can give positive assessment to Patterson's achievements, to the extent that he expounds thoroughly the relation between Tarski's 1936 article and Carnap's prior studies. As for Clue (2') and Clue (3), even though the direct relevance might be doubtful, because they refer to the years later than the period Patterson focuses on, they have been at least nicely employed indirectly (or at times implicitly) through Clue (1) and Clue (2).

2   Patterson's Interpretation of Tarski

Patterson, in Chap. 6 of Patterson (2012) entitled "Logical Consequence", is surprised by the fact that, even though Carnap's concept of consequence was the only one Tarski mentioned in his 1936 paper, virtually no one carefully studied to what extent the paper was indebted to Carnap's *Logical Syntax*. And, he claims that Tarski's article was born with the modest intention to improve Carnap's concept of consequence, and all important theses and arguments in it was targeting Carnap (Patterson 2012, p. 181). Despite its extreme boldness, I agree with Patterson's claim wholeheartedly. However, in order to appreciate the meaning of Patterson's claim, we need to locate it in the context of the whole book. In other words, we need to check first how Patterson understands the meaning of Tarski's 1936 article in the intellectual development of the young Tarski.

In fact, Patterson himself vividly presents the blueprint of his own project at the beginning of the book. According to him, the young Tarski's research before his 1936 article was all done under the influence of Intuitionistic Formalism, Though such an influence became weaker and weaker through the transition period of 1934 through 1935, it was only in 1936, when Tarski was fully freed from the influence of Intuitionistic Formalism (Patterson 2012, pp. 3–4; pp. 7–8).

In order to understand this picture, no doubt, we need to probe the questions as to (1) What is Intuitionistic Formalism?, (2) How to understand the whole early study of the young Tarski as contribution to the Intuitionistic Formalism, (3) How is it possible that Tarski's 1936 article can be the watershed sharply separating the early Tarski and the later Tarski. Interestingly, in all these three questions, the task of comparing Tarski and Carnap is playing some crucial roles in Patterson's interpretation. Carnap intervenes when we contrast Intuitionistic Formalism with

Hilbertian pure formalism.[8] Even if the early study of the young Tarski was intended to be contributions to Intuitionistic Formalism, it is quite another matter whether they can be evaluated truly as contributions to Intuitionistic Formalism, and Carnap has to intervene in this evaluation. Especially so, because it is Patterson's interpretation that Carnap sided with Leśniewski in being skeptical about semantics. If Patterson's interpretation is correct, Carnap, in at least two ways, played an important role in Tarski's divorce from Intuitionistic Formalism.

2.1 Intuitionistic Formalism and the Young Tarski

In Patterson's book, what we may count as the definition of "Intuitionistic Formalism" is presented as follows:

> "Intuitionistic Formalism" is then the view that the point of formal systems is the clear expression of thought, and that such systems are to be judged "adequate" or not based on how well their expressions, used in accord with "conventional-normative schemata", systematically accord with the representational intentions of their "speakers". (Patterson 2012, p. 23)

Also, he invokes the following five points as the most important tenets of Intuitionistic Formalism:

1. The understanding of language as a medium for the expression of thought via a speaker's intentions.
2. The conception of conventions of language as determining which thoughts are expressed by the use of a given sentence when a language is used correctly.
3. The conception of axioms and theorems as sentences that are assertible in the sense of seeming true, given the conventions, to those who are party to them.
4. The conception of the intuitive meaning or content of a term in terms of the traditional logical notion of connotation.
5. The goal, in theory construction, of constraining the interpretations of the primitives as much as possible through the implication of theorems that constrain the assignment of intuitive meaning to the terms of a theory. (Patterson 2012, p. 43)

How is the so-called Intuitionistic Formalism, understood in this way, related to the work of young Tarski? Patterson again sketches the answer in outline as follows:

> Our story, summarized, goes like this. Tarski originally became interested in the question of how our thoughts and ideas, under a certain conception of what those are, could adequately be expressed in an axiomatic theory or "deductive science", a topic that was central to his advisor Stanislaw Leśniewski's work. As Intuitionistic Formalism was conceived of by Leśniewski as a conception of the function and significance of an axiomatic theory, Tarski

---

[8]Though it is a controversial issue whether Hilbert himself was a formalist, it is simply beyond the scope of this paper to deal with it. My use of the phrase "Hilbertian pure formalism" is intentional in that it can refer to what people widely believe to be the position of Hilbert and his followers.

set himself the task of exploring how the basic concepts used in thought *about* axiomatic theories–consequence, truth, reference and related notions such as completeness and categoricity (in several senses)–could themselves be captured, to Intuitionistic Formalist standards, within an axiomatic theory. In particular due to his interest in early work in what we would now think of as model theory, Tarski set out to develop a way of capturing the semantic notions of truth, satisfaction and reference within such a theory to Intuitionistic Formalist standards. The result of this project was the now-familiar method of defining truth by recursion on satisfaction. (Patterson 2012, p. 3)

Based on Clue (1) and Clue (3), we realize that, insofar as Patterson's hypothesis that the young Tarski worked on the project intended to contribute to Intuitionistic Formalism is right, the heart of matter lies in understanding exactly the differences between the formalism of the Hilbert school and Intuitionistic Formalism.[9] Patterson introduces two distinctions in order to uncover the differences: (1) the distinction between representational semantics and expressive semantics; (2) the distinction between epistemic conception of inference and universality conception of inference. Patterson seems to understand the itinerary of Tarski from freeing himself from the influence of Intuitionistic Formalism to the establishment of his own position roughly as the transition from expressive semantics and epistemic conception of inference to representational semantics and universality conception of inference.

2.1.1 The Distinction Between Representational Semantics and Expressive Semantics

The distinction between the "representational semantics" and the "expressive semantics" is in fact introduced at the beginning part of Patterson's book. Perhaps one is bound to get the impression that it is the ultimate issue governing all problems of philosophy of language and logic, in which Tarski was interested:

This book tells the story of the birth of truth-conditional semantics from an earlier conception of meaning. If we think of language as standing between mind and world, there are two simple ways in which to think of it as meaningful: in terms of its relation to the mind, or in terms of its relation to the world. One may thus conceive of meaning in terms of the expression of thought, or in terms of the representation of things. Call the two conceptions *expressive* and *representational* semantics. On the expressive conception of the function of language is to express thoughts, which may themselves be representational. On the representational conception of language is conceived of as representational in its own right, and the expression of thought is a derivative function. Central to the expressive conception of language are the notions of assertion and justification since our basic notions are a subject's saying something and their reasons for doing so; central to the representational

---

[9]Sinaceur (2009) is another example that tries to capture the uniqueness of Tarskli's version of formalism. She claims that "the general spirit of Tarski's logico-mathematical work was formalist", and explains in what sense Tarski's work was formalist by using six points (Sinceur 2009, p. 372). Each and every of those six points is interesting. Strangely enough, however, Sinaceur does not discuss them in direct connection with intuitionistic formalism. I will return to her discussion of Tarski's intuitionistic formalism (see ibid., p. 380) later in this chapter.

conception are the notions of reference and truth being about things and accurately representing how they are. (Patterson 2012, p. 1)

It is evident that Patterson believes that the intuitionistic formalists adopted the expressive semantics:

> To return again to our contrast from the introduction, Intuitionistic Formalism is a view within expressive rather than representational semantics. (Patterson 2012, p. 40)

However, Intuitionistic Formalism is not the prime example of expressive semantics. Patterson seems to view Hilbertian pure formalism as such an example, for he presents the further details of his views as follows:

> Appeals to intuition by Leśniewski are not to be understood as having any justificatory force; the whole point of the axioms is simply to express clearly the author's thoughts. This will be important when we turn later to the question of whether the Intuitionist Formalist followed Hilbert in taking consistency to be sufficient for truth. (Patterson 2012, p. 24)

One notable fact here is that, in discussing the expressive semantics of the Intuitionistic Formalism, Patterson adopts the method of comparing the Intuitionistic Formalism with the Hilbertian pure formalism. Both are the views within expressive semantics. But Intuitionistic Formalism is not entirely like the Hilbertian pure formalism. But exactly how are they different? As we can see from the following quotes, Patterson exploits Carnap in order to answer this question.

> As distinct from a pure formalist, or the Carnap of *Logical Syntax*, Leśniewski sees difference between a deductive theory that expresses intuitive thought about reality and a merely consistent system as being of great import. (Patterson 2012, p. 26)

> Notice again the contrast here with Hilbert and especially with Carnap, whom we have seen in *Logical Syntax* taking this very problem to be the reason that one ought to ignore anything that isn't determined by the structure of a deductive theory. For Leśniewski, by contrast, the colloquial explanation, though problematic, was of central importance. (Patterson 2012, p. 29)

Carnap's position may not be entirely like the Hilbertian pure formalism either. However, according to Patterson's discussion, Carnap is closer to Hilbert than Leśniewski. Then, what exactly is unique in Leśniewski? In order to answer this question, the following quotes may be useful.

> In order to see how Intuitionistic Formalism pans out as a view about deductive sciences, we can consider, in contrast to the view we find in Carnap in Hilbert, a line of reflection on the "axiomatic method" that runs from a paper of Padoa that Tarski cites with enthusiasm, through Kotarbiński and on to remarks of Tarski that persist as late as 1937s "Sur la methode deductive" and the inclusion of the latter, with additions, in Tarski's logic textbook. This sort of view adds to the formalist one the idea that the point or function of a deductive theory, though it can be understood on its own in the formalist manner, is *to express certain intuitively valid thoughts of its creator*. (Patterson 2012, pp. 37–38. Emphasis is mine)

This contrasts with Hilbert in a way broadly in tune with Padoa: unlike Hilbert, the primitive terms of an axiom system are associated with ideas not determined by the axioms, and it is simply an intriguing sort of generality that it can turn out that the same axiom system can receive different intuitive interpretations. *The system bears an intended interpretation not settled by its structure*, though generality can be achieved by varying the interpretation imposed upon it. (Patterson 2012, p. 39. Emphasis is mine)

By the time of *Logical Syntax* there thus comes to be a major difference between Carnap and Leśniewski on what Carnap calls the "principle of tolerance". (Patterson 2012, p. 41)

In all this, Patterson seems to claim that, unlike Hilbert, (and unlike Carnap, insofar as he adopts the principle of tolerance), Leśniewski had an expressive semantics that expresses *certain intuitively valid thoughts of its creator*. As we saw above, Patterson carefully notes that "[o]n the expressive conception of the function of language is to express thoughts, which may themselves be representational" (Patterson 2012, p. 1).

2.1.2   The Distinction Between Epistemic Conception of Inference and Universality Conception of Inference

Immediately after introducing the distinction between the expressive semantics and the representational semantics, Patterson points out that the concept of inference is directly related to it:

Both views address the topic of inference. The expressive conception offers a natural, intuitive connection to the basic idea of an argument as something that gives one a reason to believe one thing given that one believes others: in a valid argument, justification for believing the premises is transmitted to the conclusion and becomes justification for believing it. The representational conception, since its focus is on accuracy in what statements are about, gives rise to the idea that an argument is valid just in case if its premises are true, so is its conclusion. Continuing in the same broad brush-strokes, the expressive conception sits naturally with the conception of inference as derivation of one claim from others in accord with intuitively valid rules—a conception that becomes, in more refined studies of logic, proof theory. The representational conception naturally leads to the idea of an inference as valid just in case all models of the premises are also models of the conclusion, an idea at the foundation of model-theoretic studies of logic. The interaction of the two conceptions in their refined forms then gives us two of the central results of the 20th century, Gödel's completeness and incompleteness theorems. (Patterson 2012, pp. 1–2)

Though we need a more detailed discussion, the big picture like this, which correlates the distinction between proof theory and the model theory with the distinction between the expressive and the representational semantics, certainly provides us with a useful perspective. Then, it becomes important to locate some leading figures the appropriate place within the big picture. Patterson points out that Leśniewski and the intuitionistic formalists did have the epistemic concept of logical consequence:

Intuitionistic Formalism centrally involved an epistemic conception of logical consequence. (Patterson 2012, p. 223)

> What is of significance for us at this juncture is simply the strongly epistemic conception of logical consequence that is at the heart of Intuitionistic Formalism as Leśniewski, is a tool for the expression of thought and reasoning. It isn't an object of study in its own right, and it must thereby be crafted to respect the logical relationships among thoughts intuitively judged. (Patterson 2012, pp. 42–43)

Given the discussion above, it is natural to expect that Patterson would connect the position of the young Tarski with the epistemic concept of consequence. In fact, Patterson claims:

> A central aspect of the doctrine to which Tarski swears fealty in 1930, then, is an epistemic conception of consequence.
>
> …, it takes Tarski a good deal of time–all the way until 1935–to venture to replace Intuitionistic Formalism's epistemic conception of inference with a generalist conception based in his semantics. (Patterson 2012, p. 18)

2.2 Skepticism About Semantics

If we are to understand the project of young Tarski in the late 1920s and the early 1930s within the framework of Intuitionistic Formalism, what did he want to achieve? We may accept Patterson's suggestion in the following quote wholeheartedly:

> Tarski's views about meaning in the late 1920s and early 1930s were entirely in line with Intuitionistic Formalism as he conceived of it following and the closely related philosophy of language of Kotarbiński. (Patterson 2012, p. 84)

However, it is by no means evident that the young Tarski's Intuitionistic Formalism was naturally led to his metamathematical project

> Tarski's project was to craft deductive theories that expressed important metamathematical concepts via the constraints placed on their primitive terms by their theorems. (Patterson 2012, ibid.)

Above of all, we do not know exactly how the young Tarski understood metamathematics. Nor do we understand what it means to say that deductive theories express metamathematical concepts. Even if we understand all this, it would be an arduous task to understand how anyone could aim at expressing metamathematical concepts by a deductive theory "via the constraints placed on their primitive terms by their theorems".

According to Hodges, Tarski took over the term 'methodology' from Lukasiewicz's 'methodology of the system'. Hodges further suggests that Tarski coined the term 'metatheory' to refer to formalized methodology, indicating its root at Hilbert's 'metamathematics' (Hodges 2008, p. 96). However, as Hodges hints at, Tarski's views about metamathematics were somewhat different from that of Hilbert's:

He claimed to be the first person to axiomatize his metatheory. (Tarski 1983, p. 173, footnote 3; Hodges 2008, p. 96)

Sinaceur's comparison of Hilbert's and Tarski's views about metamathematics seems pertinent at this stage. First, she points out the similarities between them:

He claimed in a 1930 paper that "formalized deductive disciplines form the field of research of metamathematics roughly in the same sense in which spatial entities form the field of research in geometry"* This claim of constituting metamathematics as a mathematical discipline was not fundamentally different from Hilbert's viewing *Beweistheorie* as a "new mathematics". (Sinaceur 2009, p. 370; *Tarski 1930, in Tarski 1986b, I, p. 313)

However, she quickly turns to their radical differences as follows:

But while Hilbert kept investigating mathematical-logical foundations, in order to eradicate philosophical dogmatism and eventually, to interpret Kant's *a priori* as the finite mode of thought,** Tarski did not think he was (only) contributing to the foundations of mathematics. He thought he was building a new mathematical branch on its own. (Ibid., pp. 370–371; **Hilbert 1930, in Hilbert 1935, pp. 383–385)

Now let us turn to Patterson to see how he describes the young Tarski's project under the influence of Intuitionistic Formalism in more detail:

In "The Concept of Truth in Formalized Languages" (CTFL) and related papers in the early 1930s Tarski developed the tools and techniques that are still at the center of logic, the philosophy of language and, to a lesser extent, linguistic semantics. Indeed, Tarski's achievements in this regard were so influential that today it is forgotten that representational semantics as he developed it was devised as a contribution to a certain project motivated by the expressive conception. Standard reports on Tarski's work simply treat it as an obvious early contribution to representational semantics. Lurking in the text of the classic papers, however, is a different project, one to which representational semantics was at first intended as a small contribution. Little is known about this project today other than that at one place (Tarski 1983, 62) Tarski refers to the view he was working with as "Intuitionistic Formalism". (Patterson 2012, p. 2)

This quote from Patterson is indeed quite informative for us to fathom what was going on. At least, he seems to present an interesting and promising interpretative hypothesis. Even if this is a correct answer, however, there still remains a doubt. Why did Tarski feel it necessary to pursue such a project? Though insufficient, Patterson seems to provide us a quite suggestive answer to this question by digging the hidden meaning in the quoted passage:

Tarski initially didn't think of semantics as he introduced it as a contribution to the study of meaning at all. Tarski developed semantics not as a contribution to the basic theory of meaning, but as a bit of detail work in his project of giving Intuitionistic Formalist treatments of important metamathematical concepts. In this respect, the early development of semantics is simply a bit of work that extends his work on the consequence construed derivationally around 1930. (Patterson 2012, pp. 84–85)

Then, how could such a thing happen? What does it mean to say that Tarski in introducing semantics without thinking it as a contribution to the theory of meaning? Patterson's discussion implicitly suggests that the situation in that historical period blocked Tarski for pursuing the possibility rather than that Tarski never thought about that possibility. For example, Patterson reports the following historical facts:

> Leśniewski and Carnap were openly hostile to semantic concepts, while their role in Kotarbiński's account is peripheral at most. (Patterson 2012, p. 85)

In fact, Patterson claims that Tarski's interest lies in combating with the skepticism about semantics:

> Tarski's interest in giving definitions of semantical concepts was directed toward showing the legitimacy of these early model theoretic studies in the face of the skepticism of Leśniewski and Carnap by showing how terms expressing semantic concepts could be introduced into rigorous deductive theories by definitions and in accord with the constraints imposed by Intuitionistic Formalism. (Patterson 2012, p. 85)[10]

In order to support Patterson's historical conjecture, we need a bit more solid evidence. Even if we consider it as a historical conjecture on the philosophical development of Tarski, such an interpretation of Tarski's interest might be just a projection of Patterson based on circumstantial evidences. If we consider it as a historical conjecture on the general situation of the logic community, again we need more detailed discussions. For example Patterson's following remark seems to be a contribution of such a sort.

> On the one side, we have Carnap, Leśniewski and others insisting that semantic notions simply can't play a role in serious theory; on the other, we have Skolem, the postulate theorists and others engaged in a good deal of serious theory that makes essential use of semantic notions. (Vaught 1974, 161; Patterson 2012, p. 94)[11]

Although it is merely a first step, Patterson's interpretation that, while the early study of the young Patterson can be understood as a modest attempt done within the broader framework of Intuitionistic Formalism, already was there Tarski's interest in the revolt against the skepticism about semantics should be evaluated as presenting an extremely productive and promising historical hypothesis.

---

[10]In connection with this, Patterson's discussion in the following is also extremely suggestive: "So, we have two questions about any given term in a deductive theory: are the conventions governing the assertible sentences in which it appears sufficient to express its intended connotation, and are the conventions governing them consistent with its expressing its intended connotation? The requirement that these questions be answered in the affirmative is, in embryo, the "material adequacy" requirement of Convention T./ The project, then, is to craft a deductive theory that forces, th[r]ough its theorems, its primitive terms to express the contents of certain concepts. This is what remains in Tarski's work of the appeal to connotation in Leśniewski and Kotarbiński (Patterson 2012, p. 50).

[11]Here, "the postulate theorists" refer to Veblen, Langford, and other scholars who can be lumped together under a unique tradition in early 20th century America. Interestingly, there is evidence that Tarski was influenced by this tradition even before his emigration to U.S.A. (see Scanlan 2003).

2.3   Carnap and Tarski

As we saw above, Patterson interprets Tarski's 1936 paper as announcing of his divorce from Intuitionist Formalism. If his interpretation is correct, the direct cause of such a radical change must be, as Patterson points out, "his careful reading of Carnap's *Logical Syntax of Language* and his interaction with Carnap himself" (Patterson 2012, p. 169).[12] Patterson particularly pins down the crucial point in "seeing what goes wrong in Carnap's definition of "analytic" for Language II led Tarski to see that what was missing was his semantics" [ibid.].

As Patterson points out, we are fully aware of the fact that Tarski studied Carnap (1934) very carefully, since Carnap paid tribute to Tarski for useful suggestions and corrections (Carnap 1937). Ironically, we tend to make errors here to be oblivious to the possibility that there might be some lessons Tarski learned from Carnap. With this danger in mind, let us follow Patterson's lead to reexamine the relation between Carnap and Tarski. Patterson is perceptive enough to be alert to the following fact:

> When he studied *Logical Syntax* Tarski would have found two passages that were close to his interests with respect to the topics that concern us here: the argument of §60b that truth cannot be defined in syntax, and the definitions of 'analytic' and 'contradictory' in §34. (Patterson 2012, p. 174)

What exactly did Tarski find there? As Patterson himself characterizes his project as "philosophical detective", which requires solving or resolving multiple mazes, it is simply impossible for me to report and examine his reasoning process extensively.[13] Let me just jump to Patterson's answer to the question raised above:

> So what Tarski found in §§ 60 and 34b was that, on the one hand, Carnap had no good argument against the possibility that semantic expressions be defined in syntax, and that, on the other, Carnap's own procedure with respect to Language II appealed to reference, satisfaction and truth in everything but name. The only thing preventing Carnap from seeing the fruitfulness of the truth-definitions he had nearly arrived at himself was a

---

[12]"... there was an intermediate phase, and all of the important work occurs during it: namely, a phase in which Tarski tried to blend both traditions by designing deductive theories situated within the first tradition precisely to be used for the second sort of activity" (Patterson 2012, p. 45).

[13]It should be noted that in this paper I fail to discuss some important issues such as "semantical theory of categories", "Carnap's theory of levels", and "Carnap's principle of tolerance": "... all we can conclude is that Carnap has convinced Tarski that it isn't obvious that expressions within the theory of levels don't have such meanings, contrary to Tarski's emphatic endorsement of the theory of semantical categories at (Tarski 1983, 215; Patterson 2012, p. 173); "But in the postscript we see that the basic expressive conception of language from within which his project began was losing its grip on Tarski, and that he was moving to a more open-minded, Carnapian, "tolerant" conception of alternative logical systems" (Patterson 2012, p. 174).

lingering prejudice against semantic notions backed up by the oversight of § 60 on which a properly scientific metalinguistic treatment of a language couldn't involve the terms of that language in addition to its own proper syntactic terms. (Patterson 2012, p. 178)[14]

But once Tarski set himself to thinking about what he could accomplish with his techniques for defining semantic terms, within a few months he had cast aside the old way of understanding consequence and with it, though less explicitly, the conception of meaning from which he had begun. (Patterson 2012, p. 179)

Further, Patterson presents the following observations that are directly related to clue (2) and clue (2'):

Read as a pair, ESS and CLC make an extended argument in favor of Tarskian semantics: the first article sketches the basics of how it works, while the second undoes the primary source of resistance he expected from his audience, namely, Tractarian doctrines about logic as they were found in the Vienna Circle. In particular, Tarski's account of consequence in (Tarski 2002) makes logical consequence a semantic concept in the sense of (Tarski 1983, 401), since consequence is defined in terms of models, which are defined in terms of the explicitly semantic concept of satisfaction. (Patterson 2012, p. 223)[15]

2.4 Carnap's and Tarski's Concepts of Logical Consequence

The definitions of logical consequence compared and examined in Tarski's 1936 article were (1) the young Tarski's definition by transformation rules, (2) Carnap's definition, and (3) a newly suggested definition by Tarski. Let us briefly scheme these three definitions, and see on what ground Tarski thought that (2) is superior to (1), and (3) is superior to (2).

Regarding (1), Patterson points out that the aim of Chaps. 2, 4, and 5 of Tarski (1983) was to capture the concept of logical consequence axiomatically (Patterson 2012, p. 53). As was discussed above, Tarski at the period was still understanding the concept of logical consequence as the derivation by intuitively valid rules of

---

[14]Together with this, we need to examine Patterson's discussion of what Tarski realized from Carnap (1934) as a whole without confining ourselves with sections 64 and 30. "Overall, then, Tarski's encounter with Logical Syntax allowed him to see that Carnap's treatment of analyticity really appeals to intuitive semantic notions anyway, that Carnap has no argument against the inclusion of semantic notions in serious theory, and that once Carnap's treatment of analyticity is replaced with Tarskian semantics, everything can be done without the transformation rules" (Patterson 2012, p. 179). See Carnap (1934, p. 11) for the distinction between Language I and Language II. There Carnap presented Language I as including elementary arithmetic in a restricted scope, and roughly corresponding to constructivism, finitism, and intuitionism. On the other hand, Language II includes further indeterminate concepts, and the theory of real numbers and analysis of classical mathematics. See also De Rouilhan (2009, p. 126 and p. 136).

[15]ESS and CLC refer to Chap. 15 of Tarski (1983), "The Establishment of Scientific Semantics", and Chap. 16, "On the Concept of Logical Consequence" respectively.

inference. In order to exemplify these points, Patterson quotes extensively from Tarski (1983). Let us take a look at some of them:

> An exact definition of the two concepts, of sentence and of consequence, can be given only in those branches of metamathematics in which the field of investigation is a concrete formalized discipline. On account of the generality of the present considerations, however, these concepts will here be regarded as primitive and will be characterized (*charakteriziert*) by means of a series of axioms. (Tarski 1983, pp. 30–31; Patterson 2012, p. 54)

As for the case (2), Tarski thought that the definition of consequence suggested in Carnap (1934) was too complicated and special to discuss. So, he formulated it as follows:

> The sentence *X* follows logically from the sentences of the class *K if and only if the class consisting of all the sentences of K and of the negation of X is contradictory.* (Tarski 1983, p. 414)

(3) is what Tarski suggested in his 1936 article:

> *The sentence X follows logically from the sentences of the class K if and only if every model of the class K is also a model of the sentence X.* (Tarski 1983, p. 417)

Patterson discusses whether (2) is superior to (1) as follows:

> Carnap's account of consequence is an improvement over a straightforward definition in terms of transformation rules of the sort that Tarski himself favored through 1934, and Tarski saw from *Logical Syntax* that it was headed in the right direction. He was able to see, though, as we discussed in the previous chapter, that it breaks down at just the point where semantics can be used to fix it. *The advance in Carnap's account was that by adding the "indefinite" rules of transformation, in particular the ω-rule, Carnap had a way of responding to Gödel's incompleteness theorem,* on which derivability was insufficient to capture all sentences that are true in the language of a theory if the theory is true. Since Carnap eschewed "true" in favor of "analytic", he needed a way to state Gödel's result without the forbidden term, and introducing an infinitary transformation rule allowed him to state Gödel's result as the claim that there are consequences of theories extending arithmetic that aren't derivable. (Patterson 2012, p. 187. Emphasis is mine)

Patterson reports how Tarski commented on Carnap's definition of logical consequence as follows:

> Tarski next notes Carnap's attempt at capturing the "proper concept" of logical consequence, faulting Carnap's definition for language II as "too special and complicated" (Tarski 2002, 182), his definition for language I as being unsuited for extension to "less elementary" languages (Tarski 2002, 192), and his definition in General Syntax as making the "denotation" of the defined concept "dependent in an essential way on the richness of the language which is the object of Consideration" (this is the issue with Condition (F) and Tarski's modification of the account; see below), and as depending on a basic concept of consequence in terms of stipulated transformation rules that Carnap leaves as primitive (Tarski 2002, 193). The last criticism doesn't occur in the German or, therefore, Woodger's translation; we can probably accept Hitchcock's conjecture (Stroinska and Hitchcock 2002, 158) that Tarski added it to the Polish in response to discussion at the Paris congress. However, rather than softening the criticism of Carnap, as Hitchcock suggests, the second point emphasizes a perhaps more fundamental advantage of Tarski's account, which is the

elimination of any dependence on transformation rules taken as primitively valid. This is the really revolutionary aspect of (Tarski 2002). (Patterson 2012, pp. 182–183)

So, Tarski found at least three different definitions of logical consequence in Carnap (1934), and commented on each of them, though very briefly. Among them, what Tarski himself formulated and reported was Carnap's definition for Language II. When he claimed that Carnap's definition of logical consequence for Language II is "too special and complicated", what did Tarski had in mind?

The answer to this question can be conjectured from the reason why (3) is superior to (2), i.e., what Tarski suggested about in what respects his own definition of logical consequence is superior to that of Carnap's. According to Patterson's interpretation, it seems that Tarski viewed Carnap's definition of logical consequence is "too special and complicated" insofar as it depends on the primitively valid rules of inference. In fact, Patterson finds the revolutionary aspects of Tarski's definition of logical consequence from "a perhaps more fundamental advantage of Tarski's account, which is the elimination of any dependence on transformation rules taken as primitively valid" (Patterson 2012, p. 183). Also, the following quote from Patterson can be the ground for such an understanding:

> Sometime in 1934-5 *Tarski realized that he could replace the appeal to primitively valid inference rules with a notion of consequence defined in terms of his semantics.* The result is the account of (Tarski 2002) and with that the project came to an end. (Patterson 2012, p. 8. Emphasis is mine)

> Now Carnap feigned to relinquish an epistemic conception of consequence with the principle of Tolerance, but his contrived selection of examples of languages with transformation rules that are in fact at least *prima facie* intuitively valid masks the real consequences of this. To the extent that Carnap's position is plausible, it is because of the air of intuitive validity that continues to pervade the transformation licensed in Languages I and II. Tarski, on the other hand, gives us a true generality conception of consequence. (Patterson 2012, p. 225)

> Logical consequence, on the other hand, was much slower to develop in Tarski's hands. Aside from a skeptical footnote following Gödel (Tarski 1983, 252), as late as 1934 Tarski treated logical consequence derivationally, in terms of a recognized set of apparently valid rules for asserting sentences given that others had been asserted. *Only when he realized that the notion of a model could be defined in terms of semantics, and consequence in turn defined in terms of that, did Tarski see that semantics could actually stand on its own as a treatment of language,* and at that point he moved on from the project that had originally motivated him: after 1936 Intuitionistic Formalist concerns disappear from his work and, in particular, 1944's "The Semantic Conception of Truth", though it appears to summarize the work of the 1930s, leaves out the themes characteristic of Intuitionistic Formalism. (Patterson 2012, pp. 3–4. Emphasis is mine)

As the re-evaluation of logical positivism and the early analytic philosophy in general becomes popular recently, a few notable studies have appeared on Carnap's definition of logical consequence (see De Rouilhan 2009, Awodey 2012). So, we have reason to hope for a more detailed and rigorous comparative study of Carnap and Tarski.

3   Beyond Patterson

Has Patterson's "philosophical detective work" been completed successfully? Though I have emphasized the positive aspects of Patterson's work, I have to take somewhat negative stance to this question. I do believe that we should not be reluctant to praise his achievements for opening a new horizon for seeking the historical and philosophical background and motivation for Tarski's definition of logical consequence. No doubt, he has been more successful than anyone else in this regard. Further, his questions are to the point so that his research seems to be on the right track. Nevertheless, such a positive assessment cannot and should not be led to the conclusion that he found the correct answer.

Then, where are we to find the weak spot in Patterson's study? We need to take somewhat roundabout path in order to present my views on this matter. For this purpose, let us take a look at the structure of Chap. 6 of Patterson (2012), which deals with the problem of logical consequence. I already introduced the content of the first half of Chap. 6 above. Now, we need to examine what Patterson does in the second half, and why he does such things. As we saw above, Patterson's interpretive hypothesis is that

> Tarski, in his treatment of consequence itself, intends in (Tarski 2002) only to modify Carnap's definition of L-consequence by supplying, via semantics, objectual quantification in favor of Carnap's substitutional quantification, and truth-preservation in favor of primitive rules of "direct consequence". (Patterson 2012, p. 194)

Patterson thinks that there are various sorts of supporting evidences for such a hypothesis, and actually provided three of them. I am not quite interested in examining these evidences, for I believe that Patterson's hypothesis is basically on the right track. Rather I am interested in what Patterson does in order to test his own hypothesis.

Patterson claims that the best test for the hypothesis is "in the treatment of the standard interpretive puzzle", and devotes the second half of Chap. 6 for discussing how the puzzles can be treated under the hypothesis (Patterson 2012, p. 194ff.). The puzzles refer to the central issues of the controversy developed for the past thirty years in connection with Tarski's definition of logical consequence. As is well known, Etchemendy criticized the Tarskian model-theoretic definition of logical consequence by a series of writings in the 1980s (Etchemendy 1988, 1990). Whether it was right or wrong, Etchemendy's criticism contributed enormously to deepen our historical and philosophical understanding of the concept of logical consequence by securing a plethora of objections. In his 1936 article, Tarski identified extensional adequacy and the elimination of contingency as conditions to be satisfied by the definition of logical consequence. By invoking problems of overgeneration and undergeneration, however, Etchemendy not only denies the extensional adequacy of Tarski's definition but also points out that Tarski's definition fails to eliminate contingency by treating obviously contingent things such as the size of the universe as logical through the axiom of infinity. Further, Etchemendy distinguishes between Tarski's initial analysis and his later model-theoretic analysis, and reconstructs the transition from the former to the latter in an interesting fashion. Since he criticizes both analyses, Etchemendy's criticism of Tarskian model-theoretic definition of logical consequence has played a role of the point of departure for reevaluating Tarski's achievement in the historical context of modern logic and philosophy as well as for philosophical assessment of whether Tarskian definition of logical consequence is the right one.[16] Furthermore, as Etchemendy recently published the long awaited replies to some of the important criticisms that had been raised against his theories, we might hope for the revival of interest in logical consequence (Etchemendy 2008).

What Patterson does in the second half of Chap. 6 of his book is of the character that can be subsumed under such a controversy stemming from Etchemendy: The problem of overgeneration and the problem of domain variability (Patterson 2012, pp. 194–203), the problem of modality and the so-called Tarski's fallacy (ibid., pp. 203–209), the problem of formality and the problem of logical constants (ibid., pp. 209–219). Patterson shows extreme care about the details of the complex issues, and demonstrates very persuasively that his own hypothesis can handle the puzzles involved. All this must be meaningful and remarkable. We may cite the problem of logical constants as a typical case. As is well known, at the end of his 1936 article, Tarski explicitly expressed a skeptical view about making the distinction between what is logical and what is extralogical. On the other hand, vast amount of ink has been spilt about reconstructing Tarski's positive view about making such a distinction, based on his 1966 lecture.[17] Patterson counts such an attempt to read Tarski's view in 1966 into his 1936 article as "anachronistic" (Patterson 2012, p. 211).[18]

However, I instinctively feel that something goes wrong here. Above all, the idea of testing the hypothesis that was suggested to uncover against what philosophical/historical background and motivation the young Tarski presented the definition of logical consequence by its ability to handle the puzzles that has been raised in the controversy since the late 1980s seems unnatural. We need to remember that when Etchemendy attacked Tarski's definition of logical consequence, the background and the motivation of Tarki's definition were by no means his main interest.

Now I can't help plunging into a philosophical detective work in Patterson-like fashion, by presenting a hypothesis for how and by what process Patterson arrived at his views in Patterson (2012). It seems that Patterson started his research with the content of the second half of Chap. 6, and completed his research by returning to his point of departure. It is natural that Patterson started his research from the on-going heated controversies stemming from Etchemendy and his critics. Also, it

---

[16]Park (1998) is a survey of the early stage in the controversy triggered by Etchemendy. Choi (2012) deals with more recent stages in the controversy.

[17]Tarski's 1966 lecture at the University of Buffalo was edited by Corcoran and published as Tarski (1986a). Some examples of such attempts to read Tarski's view in 1966 into his 1936 article include Corcoran and Sagüllo (2011), Feferman (1999, 2008), Gómez-Torrente (1996, 2000, 2008), Mancosu (2006, 2008, 2010a, b), Ray (1996), Sagüllo (1997), and Sher (1991, 1996, 2001, 2008). For more recent studies of similar sort, see Choi (2012).

[18]Park (1998) is a clear example of developing arguments under such an anachronistic assumption.

seems natural that he wanted to evaluate his results against the background of his starting points. What is troublesome is that, even if the second half of Chap. 6 is both points of departure and arrival, the essence of Patterson's research in Patterson (2012) regarding the problem of the definition of logical consequence is found in entirely different parts.

Patterson's research actually executed on Tarski's definition of logical consequence is no doubt authentic: starting from Tarski's 1936 article; focusing on Carnap's definition of logical consequence discussed in that article; asking in what respects Carnap's definition was superior to early Tarski's definition; asking why Tarski's definition had to be as it was; as a consequence, immersed in commentary work on some of the more relevant ones among the writings collected in Tarski (1983); asking what kind of work the young Tarski did within the framework of intuitionistic formalism; becoming curious about how, thanks to Carnap, Tarski was freed from the influence of intuitionistic formalism; arriving ultimately at the realization that (1) the distinction between representative and expressive semantics, and (2) the distinction between epistemic and generality conceptions of consequence are the guiding themes of the entire problem. I would like to praise wholeheartedly Patterson for taking such a procedure and for acquiring his handsome results. What is troubling is, however, the realization that (1) the distinction between representative and expressive semantics, and (2) the distinction between epistemic and generality conceptions of consequence are the guiding themes of the entire problem has not solved once and for all the initial problem he wanted to solve. It would be more appropriate and exact to view his outcome as discovering more difficult problems rather than solving the initial problem, for it would not be an easy matter to clarify these distinctions, not to mention uncovering the historical origins of them. It is my opinion that it would have been enough to describe such a state of affairs. After all, wouldn't that be standing on Socratic tradition of philosophy?

Presumably, the reason why Patterson tries to test his interpretive hypothesis by the recent controversies stemming from Etchemendy and his critics may be found in all too human intellectual vanity. We should not forget that in fact Patterson himself makes it explicit that the topic of his book "Tarski's evolving set of views about logic and language, and in particular "formal semantics" in the period of 1926–1936" (Patterson 2012, p. 9). Of course, there are some positive effects. Since Patterson himself presents an entirely new discussions that cannot be found in precedent studies, there is enough ground to believe that Patterson was considerate of the readers who would be interested in the relationship between his and previous studies. Perhaps, Patterson wants to hide the best clues for the scholars of the future generation who will challenge the most fundamental problem of logic and philosophy of logic still left unsolved.

References

Asmus, C., & Restall, G. (2012). A history of the consequence relations. In D. Gabbay, F. J. Pelletier, & J. Woods (Eds.), *Logic: A history of its central concepts. Handbook of the history of logic* (Vol. 11). Amsterdam: Elsevier.

Awodey, S. (2012). Explicating 'Analytic', in Wagner (2012), pp. 131–143.

Bach, C. N. (1997). Tarski's 1936 account of logical consequence. *Modern Logic, 7*(2), 109–130.

Betti, A. (2008). Polish axiomatics and its truths: On Tarski's Leśniewskian background and the Ajdukiewicz connection, in Patterson (Ed.) (2008), pp. 44–71.

Carnap, R. (1934, 1937). *The logical syntax of language.* London: RKP.

Carnap, R. (1963). Intellectual autobiography. In Schlipp, P. (ed.), *The Philosophy of Rudolf Carnap.* LaSalle:Open Court, pp. 3–84.

Choi, W. (2012). Model-theoretic consequence and modality. *The Korean Journal for History of Mathematics, 25*(4), 21–36 (in Korean).

Corcoran, J., & Sagüillo, J. M. (2011). The absence of multiple universes of discourse in the 1936 Tarski consequence-definition paper. *History and Philosophy of Logic, 32,* 359–374.

De Rouilhan, P. (2009). Carnap on logical consequence for languages I and II, in Wagner (2009), pp. 121–146.

Etchemendy, J. (1988). Tarski on truth and logical consequence. *Journal of Symbolic Logic, 53,* 51–79.

Etchemendy, J. (1990). *The concept of logical consequence.* Cambridge, MA: Harvard U.P.

Etchemendy, J. (2008). Reflection on consequence, in Patterson, D. (2008), pp. 263–299.

Feferman, A. B., & Feferman, S. (2004). *Alfred Tarski: life and logic.* Cambridge: Cambridge University Press.

Feferman, S. (1999). Tarski and Goedel: Between the lines. In *Alfred Tarski and the Vienna Circle, Vienna, 1997* (pp. 53–63). Dordrecht.

Feferman, S. (2008). Tarski's conceptual analysis of semantic notions, in Patterson (2008), pp. 72–93.

Frost-Arnold, G. (2013), *Carnap, Tarski, and quine's year together: Conversations on logic, math, and science* (Monograph and edited translation). Open Court Press.

Gómez-Torrente, M. (1996). Tarski on logical consequence. *Notre Dame Journal of Formal Logic, 37,* 125–151.

Gómez-Torrente, M. (2000). A note on formality and logical consequence. *Journal of Philosophical Logic, 29,* 529–539.

Gómez-Torrente, M. (2008). Are there model-theoretic logical truths that are not logically true? In D. Patterson (Ed.), *New essays on Tarski and philosophy* (pp. 340–368). Oxford: Oxford University Press.

Hilbert, D. (1930). Naturerkennen und Logik. *Naturwissenschaften, 18,* 959–963, in Hilbert (1935), pp. 378–387.

Hilbert, D. (1935). *Gesammelte Abhandlungen* III, Berlin.

Hodges, W. (2008). Tarski's theory of definition, in Patterson (Ed.) (2008), pp. 94–132.

Jané, I. (2006). What Is Tarski's Common Concept of Consequence?. *The Bulletin of Symbolic Logic, 12* (1),1–42.

Mancosu, P. (2006). Tarski on models and logical consequence. In J. Ferreirós & J. J. Gray (Eds.), *The architecture of modern mathematics* (pp. 209–237). Oxford: Oxford University Press.

Mancosu, P. (2008). Tarski, Neurath and Kokoszynska on the semantic conception of truth, in Patterson (2008), pp. 192–224.

Mancosu, P. (2010a). Fixed-versus variable-domain interprestations of Tarski's account of logical consequence. *Philosophy Compass, 5*(9), 745–759.

Mancosu, P. (2010b). *The adventure of reason. Interplay between mathematical logic and philosophy of mathematics: 1900-1940.* Oxford: Oxford University Press.

Park, W. (1998). On what is logical and what is extralogical. *Korean Journal of Logic, 2*, 7–32 (in Korean).

Park, W. (2014). The historical background of Tarski's definition of logical consequence. *Korean Journal of Logic, 17*(1), 33–69.

Patterson, D. (Ed.). (2008). *New essays on Tarski and philosophy*. Oxford: Oxford University Press.

Patterson, D. (2012). *Alfred Tarski: Philosophy of language and logic*. Palgrave Macmillan.

Ray, G. (1996). Logical consequence: A defense of Tarski. *Journal of Philosophical Logic, 25*(6), 617–677.

Sagüillo, J. M. (1997). Logical consequence revisited. *Bulletin of Symbolic Logic, 3*(2), 216–241.

Scanlan, M. (2003). American postulate theorists and Alfred Tarski. *History and Philosophy of Logic, 24*, 307–325.

Sher, G. (1991). *The bounds of logic: A generalized viewpoint*. Cambridge: MIT.

Sher, G. (1996). Did Tarski commit 'Tarski's fallacy'? *Journal of Symbolic Logic, 61*(2), 653–686.

Sher, G. (2001). The formal-structural view of logical consequence. *The Philosophical Review, 110*, 241–261.

Sher, G. (2008). "Tarski's Thesis", in Patterson (2008), pp. 300–339.

Sinaceur, H. (2009). Tarski's practice and philosophy: Between formalism and pragmatism. In S. Lindström, et al. (Eds.), *Logicism, intuitionism, and formalism: What has become of them?* (pp. 357–396). Boston: Springer.

Stroinska, M., & Hitchcock, D. (2002). Introduction, in Tarski, A. (2002), pp. 155–175.

Sundholm, G. (2003). Tarski and Leśniewski on languages with meaning versus languages without use. In: J. Hintikka et al. (Eds.), *Philosophy and logic: In search of the polish tradition* (pp. 09–128). Dordrecht: Kluwer.

Tarski, A. (1930). "Über einige fundamentale Begriffe der Metamathematik. *Compte Rendus de la Société des Sciences et des Lettres de Varsovie, XXIII*, Cl. 3, 22–29. Reprinted in Tarski (1986b), *I*, 311–320. English translation in Tarski (1983), 30–37.

Tarski, A. (1936a). On the concept of logical consequence, in Tarski (1983), pp. 409–420.

Tarski, A. (1936b). The establishment of scientific semantics, in Tarski (1983), pp. 401–408.

Tarski, A. et al. (1971) Undecidable theories, in collaboration with Andrzej Mostowski and Raphael M.Robinson, North-Holland, Amsterdam.

Tarski, A. (1983). *Logic, semantics, metamathematics* (J. H. Woodger, Trans.) (2 ed. J. Corcoran, Ed.). Indianapolis, Indiana: Hackett Publishing Company.

Tarski, A. (1986a). What are logical notions? *History and Philosophy of Logic, 7*, 143–154.

Tarski, A. (1986b). *Collected papers*, I–IV (S. R. Givant, & R. N. McKenzie, eds.). Birkhäuser.

Tarski, A. (2002). "On the concept of following logically", translated from the Polish and German by Magda Stroinska and David Hitchcock. *History and Philosophy of Logic, 23*, 155–196.

Tarski, A., & Vaught, R. L. (1957). Arithmetical extensions of relational systems. *Compositio Mathematica, 13*, 81–102.

Vaught, R. L. (1974). Model theory before 1945, in Henkin L. et al. (1974), pp. 153–186.

Vaught, R. L. (1986). Alfred Tarski's work in model theory. *Journal of Symbolic Logic, 51*(4), 869–882.

Woleński, J. (1989). Brentano's criticism of the correspondence conception of truth and Tarski's semantic theory. *Topoi, 8*(2), 105–110.

Woleński, J. (1995). On Tarski's background. In J. Hintikka (Ed.), *From dedekind to Gödel* (pp. 331–341). Boston, MA, 1992 (Dordrecht).

Woleński, J. (1999). Semantic revolution-Rudolf Carnap, Kurt Gödel, Alfred Tarski. *Alfred Tarski and the Vienna Circle* (pp. 1–15). Vienna, Dordrecht: Kluwer.

Chapter 9
On the Motivations of Gödel's Ontological Proof

Abstract In recent years there has been a surge of interest in Gödel's ontological proof of the existence of God. In spite of all this extensive concern, it is not certain whether there is any improvement in understanding the motivations of Gödel's ontological proof. Why was Gödel so preoccupied with completing his own onto-logical proof? To the best of my knowledge, no one has dealt with this basic question seriously enough to answer it. In this chapter, I propose to examine Gödel's ideas against a somewhat larger background in order to understand his motivation for establishing the ontological proof. I shall point out that the value of Gödel's proof is to be found in the possible role of his proof of the existence of God in his philosophy as a whole as well as in its relative merit as an ontological proof. Hopefully, my guiding question as to Gödel's motivation will turn out to be extremely fruitful by enabling us to fathom his mind regarding God and mathematics.

Keywords Axiomatic method · Existence of god · Gödel · Ontological proof

1 Introduction

In recent years there has been a surge of interest in Gödel's ontological proof of the existence of God. Gödel showed his proof (Gödel *1970: Godel 1995, pp. 388f.; See also Adams 1995) to Scott, and Scott made a note of the proof and presented it in his seminar at Princeton University in the fall of 1970. From then on, Gödel's proof has become widely circulated. It was finally published in Sobel (1987) as an appendix and later included in volume three of Gödel's *Collected Works*. Recent discussions of Gödel's proof mostly start from Sobel's crticisms. As is well known, the most influential criticism of Sobel is that Gödel's proof leads to a consequence unacceptable to most philosophers, i.e. that all truths are necessary truths. Anderson (1990) viewed this as the modal collapse of Gödel's assumptions, and tried to save Gödel's proof by some plausible modifications. Anderson's emendation secured

This chapter was originally published as Park (2003).

© Springer International Publishing AG, part of Springer Nature 2018
W. Park, *Philosophy's Loss of Logic to Mathematics*, Studies in Applied Philosophy, Epistemology and Rational Ethics 43, https://doi.org/10.1007/978-3-319-95147-8_9

many interesting responses including Oppy (1996), where a parody of the Gödelian proof reminiscent of Gaunilo's objection to Anselm's proof is presented. As one might expect, such a parody has invited friends of ontological proofs to follow in the footsteps of Anselm.

In spite of all this extensive concern, it is not certain whether there is any improvement in understanding the motivations of Gödel's ontological proof. Why was Gödel so preoccupied with completing his own ontological proof? To the best of my knowledge, no one has dealt with this basic question seriously enough to answer it.

In this chapter, I propose to examine Gödel's ideas against a somewhat larger background in order to understand his motivation for establishing the ontological proof. I shall point out that the value of Gödel's proof is to be found in the possible role of his proof of the existence of God in his philosophy as a whole as well as in its relative merit as an ontological proof. Hopefully, my guiding question as to Gödel's motivation will turn out to be extremely fruitful by enabling us to fathom his mind regarding God and mathematics.

2   Back to Gödel's Original Proof

It should be noted that what Sobel discussed was not the proof given in Gödel's own writing but the proof in Scott's note. Sobel did so on the assumption that the ideas presented in Scott's note are those of Gödel's substantially agreeing with "ideas conveyed in two pages of notes in Gödel's own hand dated 10 February 1970 and entitled "Ontologischer Beweis", which appears in *Collected Works* Vol. III as *1970. To say the least, it is not certain whether such a treatment of Gödel's proof is legitimate. In the appendix, Sobel juxtaposed both Gödel's version and that of Scott's. At a glance, they are significantly different. For example, Gödel's axiom 1 is missing or transformed into axiom 2 in Scott's version. It is left as homework for the reader to check their comprehensional, intensional, extensional (or what not) equivalence.

Anderson heartily accepted Sobel's announcement of the modal collapse of Scott's version of Gödel's ontological proof. But with slight modification, he thought, the collapse can be avoided. So, we have Anderson's emendation of Scott's version of Gödel's ontological proof. Now Oppy freely identified Anderson's emendation of Scott's version of Gödel's ontological proof as Gödel's proof and parodied it to hail Gaunilist victory. What is going on?

Presumably Anderson is one of the best friends of ontological proofs, and in view of the damaging effect of the charge of modal collapse, his alleged emendation must be an example of heroic benevolence to save non-existent Gödel. But was emendation needed at all? In his introductory note to Gödel *1970, Adams expressed exactly that kind of suspicion:

> It is characteristic of Leibnizian philosophical theology to be in some danger of leaving no truths contingent (see Adams 1977). And it is not altogether clear that Gödel was determined to avoid such a necessitarian conclusion. (See also Dawson 1997, p. 266)

It is widely known that Leibniz was one of the intellectual heroes of Gödel. Further, as Adams noted, "The study of Leibniz is known to have been a major intellectual preoccupation for Gödel during the 1930s (Menger 1981, §§ 8, 12) and especially during 1943–46 (Wang 1987, pp. 19, 21, 27)". We can easily pile up more testimonies to advocate Gödel's Leibniz scholarship. Also, it has been duly noted that Gödel's ontological proof is basically Leibnizian. Adams seems to be primarily interested in the Leibnizian character of Gödel's ontological proof in such a way that he organized his introductory note to Gödel's ontological proof on the premise that at least Gödel knew pretty well both that "Leibniz held that Descartes's ontological proof is incomplete" and that "Leibniz also held that the ontological proof can be completed by proving the possibility of God's existence".

There seems to be, then, reasonable doubt as to whether Gödel would have been impaled by the threat of modal collapse. If so, Anderson's kindness could have been either misplaced or presumptuous on Gödel's part. Needless to say, the original must be consulted if possible. In this case, Gödel's original proof in perfect shape. There seems to be no reason why one should discuss allegedly Gödelian proofs rather than Gödel's original proof.

One more reason to go back to Gödel's original proof is this. Though cryptic, ample clues are found for fathoming his mind in the philosophical annotations in *1970. Gödel's *Collected Works* also contain "Texts relating to the ontological proof" as Appendix B, where more clues are found. Curiously enough, commentators of Gödel's ontological proof have not in general fully utilized these clues. Wang (1996) seems the only case where Gödel's annotations are discussed in detail. Unfortunately, Wang was not sympathetic at all to Gödel's interest in the ontological proof (see Wang 1988, p. 195, 1996, p. 121). Be that as it may, he failed to treat Gödel's ontological proof in close connection with Gödel's views on other philosophical issues, in particular on the philosophy of mathematics. Apparently, commentators of Gödel's ontological proof did not treat Gödel seriously as a professional philosopher let alone as a historian of ontological proofs or a Leibniz expert. As a consequence, they have failed to discuss some of the most enlightening clues found in Gödel's writing. At least, so I shall argue.

3 Gathering Clues

I would like to gather as many clues as possible for understanding Gödel's motivation for drafting his ontological proof from his philosophical annotations in *1970 and from the "Texts relating to the ontological proof".

(1) **Axiomatic method**

The most prominent aspect of Gödel's ontological proof must be that it utilizes the axiomatic method. There may have been predecessors in the history of ontological arguments, but let us not forget the fact that we are dealing with Gödel's axiomatic method in the post-Hilbertian era. Gödel's proof has axioms and definitions, and if Gödel has his own view of the roles of axioms and definitions as a leading mathematical logician, that must be relevant to understanding his ontological proof.

In Excerpt from "Phil XIV", we read:

> *Philosophy*: The ontological proof must be grounded on the concept of value (p better than $\sim p$) and on the axioms:[X]

In the footnote[X] Gödel elaborated the point: "It can be grounded only on *axioms* and *not* on a definition (=construction) of "positive," for a construction is compatible with an arbitrary relationship."

These remarks provide us with at least two hints for further query. First, if we remember the question as to whether "that than which nothing greater can be thought" in Anselm's proof was meant to be a definition of God, then here we might have some data about Gödel's view on Anselm's ontological proof. But do we have any axioms in Anselm's proof in *proslogion*? Even if it has some, do they play the role as Gödel would assign to axioms? If there is none, then Gödel already has refuted Anselm's proof. We may generalize the point to all previous ontological proofs to check whether all of them are vulnerable to Gödel's (implicit) criticism.

Secondly, though inseparable from the first point, we have to examine Gödel's remark here to see whether he was criticizing Hilbert's view regarding implicit definitions and other issues of axiomatic method. No one has shed light on Gödel's use of the axiomatic method in *1970 by fully exploiting Gödel's views on the foundations of logic and mathematics.

(2) Why ontological proof?

However, probably the most relevant to our guiding question is found in Excerpts from "Max XI". In the item [From page 97] we read:

> Remark(Philosophy): If the ontological proof is correct, then one can obtain insight a priori into the existence (actuality) of a non-conceptual object.

But what is securing a priori the existence (actuality) of a non-conceptual object for? Except as a possible proof of platonism in mathematics and mathematical intuition, I cannot figure out any other use. But let us postpone this issue for a while.

Another possible reading of "Why ontological proof?" is to interpret the question as meaning "Why not cosmological proof but ontological proof?" Though some would view this kind of approach as reading too much history of ontological and other traditional arguments in favor of God's existence into Gödel, the mathematician and amateur philosopher, such a qualm would have no ground. For in the item [From page 149] we read the following remarkable passage:

> Remark (Theology): The reflection: according to the Principle of Sufficient Reason the world must have a cause. This must be necessary in itself (otherwise it would require a further cause). Proof of the existence of an a priori proof of the existence of God (the proof it contains fails to be one).

This passage is remarkable since it is a rare example of trying to understand the relationship between cosmological and ontological proofs. Why did Aquinas adopt cosmological proofs rather than ontological proofs in his five ways? Why did Scotus or Descartes attempt ontological proof again with full understanding of Aquinas's attack? Regardless of the intrinsic value of questions like these, Gödel must have been struggling with the similar questions. Intuitively, I tend to interpret his reflection as saying that cosmological proof is merely a *quia* demonstration (demonstration of fact) not a *propter quid* demonstration (demonstration of reasoned fact) (see Mancosu 1996, pp. 11–13). What is nice in this particular reading is that it reminds one of the plausible purpose of seeking ontological proofs. Anselm is a perfect example. Immediately before he presented his celebrated ontological proof, Anselm wanted to make sure that the sole purpose lies in the attitude of "faith seeking understanding". As a believer, he already had faith in the existence of God. But he wanted to understand God by ontological proof. Gödel seems to be on a par with Anselm in this respect.

Of course, it is difficult to make sense of the final part of the passage. If I am right, Gödel is saying that cosmological proof itself cannot be the proof of the existence of an a priori proof of the existence of God. Probably, the only possible way of providing such a proof of the existence of an a priori proof of the existence of God is to deliver a priori proof of the existence of God. So, Gödel tried to give the ontological proof.

(3) Positive, privation, etc.

Another important clue for in depth understanding of Gödel's proof must be found in what Gödel meant by "positive". In fact, Gödel was anxious to comment on it in *1970 itself:

> Positive means positive in the moral aesthetic sense (independent of the accidental structure of the world). Only then [are] the axioms true. It may also mean pure "attribution" as opposed to "privation" (or containing privation). This interpretation [supports a] simpler proof.

Commentators of Gödel's ontological proof have been prudent enough not to indulge in far-fetched speculation about what Gödel meant by "positive". At best, they alluded to simple or atomic propositions. But if that is all Gödel meant, why did he invoke "moral" and "aesthetic"? Again Wang (1996) seems to be the only case that contains some useful, though hardly satisfactory, discussion of the possible meaning of "positive". Speculation is needed here! My hunch is that Gödel is deeply indebted to the medieval theory of transcendentals, which was rooted in neoplatonism. In the context of discussing ontological proof, being and truth were already at issue, and Gödel seems to remind us of the convertibility of transcendentals by introducing the moral aesthetic sense of "positive". Wang (1996, p. 120) has interesting comments possibly supporting my hunch: "Gödel seems to identify the true with the good (and the beautiful). The affirmation of being is both the cause and the purpose of the world." Mentioning "privation" seems perfectly consistent with interpreting Gödel as a neoplatonist. The item entitled "*Philosophy*:

Ontological Proof" in Excerpt from "Phil XIV" contains three possible interpretations of "positive" and other more pregnant information for our subject. It is simply impossible for me to discuss them now. Let it suffice to note that these remarks further demonstrate how seriously Gödel was involved in uncovering the meaning of "positive", and that there is no counter evidence against viewing Gödel as a neoplatonist.

(4) Cause, analyticity, etc.

Still another important clue we cannot ignore in "Texts relating to the ontological proof" is related to the concept of cause, analyticity, the relationship between mathematics and physics, and other cognate issues. In the items entitled *"Philosophy"* in Excerpt from "Phil XIV", Gödel declares:

> The fundamental philosophical concept is cause. It involves: will,[x] force, enjoyment,[x] God, time, space.[*] ([*]Being near = possibility of influence. [x]Hence life and affirmation and negation.). (pp. 433–434) (see Wang 1996, p. 120)

Later in the same item he wrote:

> Perhaps the other Kantian categories (that is, the logical [categories], including necessity) can be defined in terms of causality, and the logical (set-theoretical) axioms can be derived from the axioms for causality. [Property = cause of the difference of things]. Moreover, it should be expected that analytical mechanics would follow from such an axiom. (pp. 433–434)

Finally, in the item entitled "Philosophy:?", we read:

> The only synthetic propositions are those of the form (a) (for example: I have this property), for these have no objective meaning, or: They depend not on God, but on the thing a. (p. 437)

What is revealing in lumping together all these remarks is that we can get a glimpse of Gödel's architectonic of sciences. The idea that analytical mechanics would follow from logical (set-theoretical) axioms, and thereby ultimately from axioms of causality alone deserves extensive philosophical discussion. It is indeed a radical vision encompassing all human knowledge on a hierarchical system. Wang reports that Gödel once told him that "there is a sense of cause according to which axioms cause theorems" (Wang 1996, p. 120). This can be a link with the clue we secured in connection with the axiomatic method.

4   The Role of the Proof of the Existence of God in Mathematics and Science

In the previous section we have gathered many suggestive clues from *1979 and "Texts relating to the ontological proof". With all these clues we can more efficiently look into Gödel's other writings in *Collected Works* or from biographies of Gödel. In order not to be led astray by the formidable data, I would like to exploit one working hypothesis: Gödel's preoccupation with the ontological proof must be found somewhere between God and mathematics. We should ask what possible role the ontological proof was supposed to play in the Gödelian edifice of human knowledge including mathematics and science.

After all, Gödel is one of the greatest mathematicians. This obvious but all too easily forgotten point leads one to realize why Gödel's ontological proof does matter. It does matter simply because it is Gödel's proof. Gödel is above all a mathematician. Probably no further justification is necessary for my choice of the working hypothesis. I would also say that all the clues are pointing to that hypothesis. They are not unfamiliar to us simply because they are relevant to the same old issues extensively discussed by commentators of Gödel's philosophy of mathematics.

Now, with all the clues and the working hypothesis, the most pertinent texts we have to turn to seem to be *1951, *1953/9, and *1961/? in vol. III of *Collected Works*. In particular, *1951 seems the best for my purpose. The key issues are centering around Hilbert Program, Carnap, platonism in mathematics, and the relationship between mathematics and physics. I will focus on how much light the clues from *1970 and "Texts relating to the ontological proof" shed on Gödel's philosophy of mathematics by raising the following two questions: (1) Does Gödel's ontological proof improve our understanding of the philosophical implications of his incompleteness result?; (2) Does it have a role in his criticism of Carnap's conventionalism in logic and mathematics, and thereby in his defense (or proof) of platonism in mathematics?

The answer to the first question seems to be "Yes"! Let us take a look at a pertinent text from *1951 where Gödel presents an argument to the effect that his second incompleteness theorem makes the incompletability (or inexhaustibility) of mathematics evident:

> It is *this* theorem which makes the incompletability of mathematics particularly evident. For, *it makes it impossible that someone should set up a certain well-defined system of axioms and rules and consistently make the following assertion about it: All of these axioms and rules I perceive (with mathematical certitude) to be correct, and moreover I believe that they contain all of mathematics.* If someone makes such a statement he contradicts himself. For if he perceives the axioms under consideration to be correct, he also perceives (with the same certainty) that they are consistent. Hence he has a mathematical insight not derivable from his axioms. (10–11; p. 309)

How should we understand what Gödel meant by the incompletability of mathematics? He himself is here cautious enough to guard against possible misunderstanding. He wrote:

> Does it mean that no well-defined system of correct axioms can contain all of mathematics proper? It does, if by mathematics proper is understood the system of all true mathematical propositions; it does not, however, if one understands by the system all demonstrable mathematical propositions. (11)

He calls the two different senses "mathematics in the objective sense" and "in the subjective sense" respectively. If the enterprise of ontological proof is successful, we may freely talk about mathematics from God's point of view. That must be nothing but mathematics in the objective sense.

If so, it is hard not to remember Aquinas's distinction between "in itself" and "to us". In launching his Five Ways, Aquinas dealt with three questions consecutively: (i) Is it self-evident that there is God?; (ii) Can it be made evident?; and (iii) Is there God?. Bluntly speaking, his idea is that even if God's existence is self-evident, still there is need for the proof. For it is not evident to us. But it is possible to make it evident to us by his Five Ways. I think, we may safely assume that Gödel was at least well versed with Aquinas's Five Ways to the extent of adopting a useful distinction. Just as Aquinas had to prove God's existence, which in itself is self-evident but not to us, Gödel had to prove the second incompleteness theorem, which is true in mathematics in the objective sense but not in the subjective sense. If this analogy is well taken, then we can understand how important the ontological proof was to Gödel. Without the solid distinction between mathematics in the objective sense and in the subjective sense, his project of proving his theorem would never get off the ground.

If the first question gets the positive answer, so does the second question. For we are already well informed of the fact that Gödel extensively used his incompleteness results in his attack on conventionalism. In his introductory note to *1953/9 Goldfarb even claims that "[t]he heart of Gödel's criticism is an argument based on his Second Incompleteness Theorem." (p. 327) So, if ontological proof is instrumental for understanding his incompleteness theorem, it must be valuable for understanding his defense of platonism and his criticism of conventionalism. Let us again look into *1951 in order to confirm this point. As is well-known, Gödel was anxious to refute the view that mathematics is our own creation. For that very reason, in refuting that view, he had to reveal his meaning of creation and creator. For example, Gödel wrote:

> On the other hand, the second alternative, where there exist absolutely undecidable mathematical propositions, seems to disprove the view that mathematics is only our own creation; for the creator necessarily knows all properties of his creatures, because they can't have any others except those he has given to them. (15–16; p. 311)

Gödel's proof of God's omniscience from creature's total dependence on God is in itself interesting. And it becomes more interesting if we remember the total dependence of painting on the painter in Anselm's proof. In Anselm's proof, the painter already had the complete idea of the painting. The painting cannot have any other properties except those the painter has given to them. Whether Gödel himself had some such a picture in mind, the parallel seems striking. It could be confusing whether the concept of God as Artist does help the philosophy of art or the concept of creation in art can help us sharpen our view of creation. An analogous situation seems to be involved in Gödel's two great projects: incompleteness proof and the ontological proof of the existence of God. Be that as it may, it is tempting to speculate whether Gödel was already helped by his idea of ontological proof in achieving his incompleteness result. If it is indeed the case, then we may speculate even further that that was the reason for Gödel's persistent preoccupation with the ontological proof. Didn't Gödel want to solve some extremely difficult problem by sharpening his idea of God? As was pointed out, for Gödel, improving the ontological proof by searching for the appropriate axioms was nothing but the way to understand God.

Since we are dealing with Gödel's platonism, it would not be inappropriate to reinvoke our clues, especially the possibility that Gödel was a neoplatonist. In *1951, there is an interesting passage which reminds us of Augustine's criticism of Academician skepticism. Gödel wanted to differentiate his own concept of analyticity from the usual logical positivist concept of "truth owing to our definitions". "Analytic" means to him "rather true owing to the nature of the concepts occurring [therein]", in contradistinction to "true owing to the properties and the behaviour of things." (34; p. 321) In that context, Gödel wants to protest against the frequent allegation of the paradoxes of set theory as a disproof of platonism. He counts that as unjust on the ground of an analogy with visual perception:

> Our visual perceptions sometimes contradict our tactile perceptions, for example, in the case of a rod immersed in water, but nobody in his right mind will conclude from this fact that the outer world does not exist. (34) (Augustine 1995, p. 75)

In Augustine's *Against the Academicians*, 3.11.26 we read:

> Surely it's the truth! There is a cause intervening so that the oar should seem bent. If it were to appear straight while dipped in the water, then with good reason I would blame my eyes for giving a false report.

5 Conclusion

Despite the recent surge of interest in Gödel's ontological proof, the level of understanding concerning it is incredibly low in that we are still ignorant of why Gödel was preoccupied with it. In this chapter, I tried to show that by probing its role in Gödel's philosophy as a whole or his philosophy of mathematics in particular we can improve our understanding of both. One of the lessons is the realization that Gödel had ample historical knowledge of philosophy in general and the ontological proofs in particular.

References

Adams, R. M. (1977). Leibniz's Theories of Contingency. *Rice University Studies, 63* (4), 1–41.
Adams, R. M. (1995). Introductory note to *1970. *Gödel, 1995*, 388–402.
Anderson, C. A. (1990). Some emendations on Gödel's ontological proof. *Faith and Philosophy, 7*, 291–303.
Augustine. (1995). *Against the academicians and the teacher* (translated with Introduction and Notes by P. King). Indianapolis: Hackett.

Dawson, J. W., Jr. (1997). *Logical Dilemmas: The life and work of Kurt Gödel*. Wellesley, MA: AK Peters.

Gödel, K. (1995). Collected works. In S. Feferman (editor-in-chief), *Unpublished essays and lectures* (Vol. III). New York: Oxford University Press.

Mancosu, P. (1996). *Philosophy of mathematics and mathematical practice in the seventeenth century*. New York: Oxford University Press.

Menger, K. (1981). *Recollections of Kurt Godel, private memoir, to appear in Schimanovich et al. 199? Wahrheitund Beweisheit*. Leben und Werk Kurt Godels. Vienna: Holer-Pichler-Tempsky.

Oppy, G. (1996). Gödelian ontological arguments. *Analysis, 56*(4), 220–230.

Park, W. (2003). On the motivations of Goedel's ontological proof. *Modern Schoolman, 80,* 144–153.

Sobel, J. H. (1987). Gödel's ontological proof. In J. J. Thomson (Ed.), *On being and saying: Essays for Richard Cartwright* (pp. 241–261). Cambridge, MA: MIT Press.

Wang, H. (1996). *A logical journey: From Gödel to philosophy*. Cambridge, MA: The MIT Press.

Wang, H. (1988). *Reflections on Kurt Gödel*. Cambridge, MA: The MIT Press.

Part IV Back to Aristotle

Chapter 10
Ontological Regress of Maddy's Mathematical Naturalism

Abstract This chapter is an attempt to probe the question as to why Maddy gave up mathematical realism and moved to her own version of mathematical naturalism. According to one widespread hypothesis, Maddy's change of mind was brought up by her criticism of Quine-Putnam indispensability argument. Though quite convincing, it is not good enough to explain why one has to give up mathematical realism. The analogy of science and mathematics will instead be shown to be the better perspective to fathom Maddy's changing beliefs. For this purpose, we have to understand to what extent Maddy's thought in her realist years, which was strongly influenced by Quine and Gödel, was governed by the analogy between science and mathematics. Also, we have to understand why Maddy gave up the analogy, and thereby gave up mathematical realism. Finally, some criticisms against Maddy's abandonment of the analogy will be examined so as to hint at the reasons why I believe Maddy's intellectual journey in mathematical ontology is rather regress than progress.

Keywords Mathematical realism · Mathematical naturalism · Maddy
Indispensability argument · Analogy of science and mathematics
Quine · Gödel

1 Introduction

Philosophy of mathematics without history and practice of mathematics sounds absurd. However, due to the rapid and far-reaching progress of mathematics in the 19th and the 20th centuries, it becomes extremely difficult, if not impossible, to expect solid background and deep understanding of mathematics from philosophers. Also, even the foundations of mathematics, which has influenced mathematics enormously in the early twentieth century, and has established itself as a part of mathematics by now, seems to be alienated from the practicing mathematicians

An earlier version of his chapter was published as Park (2006) in Korean.

nowadays. All these facts seem to go together. The reason why we need to focus on Penelope Maddy among the many distinguished philosophers of mathematics in our time thus should be found in the fact that she has worked very closely with the practicing mathematicians, thereby representing their positions pretty well in her writings. What motivates this chapter is the idea of examining critically Maddy's ontological development without giving up her basic orientation in mathematical naturalism throughout all these years at the center of the controversies in philosophy of mathematics.

To be sure, most of the practicing mathematicians could be counted as realists. Even though the mathematical objects studied by mathematicians are abstract entities that do not actually exist in physical space and time, (and, of course, they would escape to some kind of formalism when faced with incisive philosophical questions) they could be characterized as "unreflective platonists" in that they do have any interest in the justification of their existence. Though Maddy as a mathematical naturalist finds no fault with such attitudes of mathematicians, she as a philosopher of mathematics was not able to avoid Benacerraf's ontological and epistemological challenges in the 1960s and 1970s against Platonism Maddy's book *Realism in Mathematics* (1990) shows Maddy's itinerary from a unreflective Platonist to a weakened realist position, which amounts to a virtual Aristotelian, in responding to Benacerraf's challenge. Nevertheless, it is important to note that she was still an firm advocate of realism. Since then, however, the keyword for her writings has been "naturalism". The result of her continued study of the recent history of axiomatic set theory, and her critical examination of Quine and Gödel, who were both unique Platonists, has been synthesized in her book *Naturalism in Mathematics* (1997), in which she finally declares the abandonment of realism. Meanwhile, Maddy's more recent studies characterizes her own work after the abandonment of the weakened realism as arealism, thereby differentiating her position from other naturalistic philosophers.

Why did Maddy turn to a position that could be counted as a sort of anti-realism by giving up realism under the name of naturalism instead of sophisticating mathematical Aristotelianism in line with her early studies? No doubt raising this question largely depends on my preconceived ideas. Maddy shows the appearance that she discusses the problem of the boundaries between natural science, mathematics, and philosophy by resorting to Gödel at crucial moments, since she, unlike Quine, emphasizes the differences between the methods in mathematics and natural science. Nevertheless, I have serious doubts as to whether she merely repeatedly emphasizes her belief that whenever there is a conflict between philosophy and mathematics or science, philosophy should succumb to them. Starting from such a doubt, I desire to argue for characterizing the development of Maddy's ontological positions, is contrary to her thought, virtual regress. No more could be expected, if I can reveal at the same time how Maddy distorts Gödel's position in subtle ways. However, in order for such an assessment to be persuasive, it must be a prerequisite to understand and explain why she had to change her position. Morerover, such a work should uncover exactly what changes were made and in what manner. If we merely summarize the change as "from realism to naturalism", we might run the risk of missing several important points. For that reason, naturalism is particularly important, for (as we mentioned above), naturalistic tendency was already there in Maddy's realist years. Also, since naturalism itself cannot be viewed as an ontological position that can rival realism, a bit more detailed explanation of the significance of her changes is needed, at least for avoiding confusion and misunderstanding.

My strategy is as follows. In Sect. 2, I shall present Maddy's position in her realist years centering around *Realism in Mathematics*. Above all, I shall show how Maddy responded to Benacerraf's epistemological and ontological challenges to platonism, in order to understand Maddy's position against the background of the problem situation at that time. Then, I shall try to grasp Maddy's position by discussing her indebtedness to Quine's and Gödel's Platonisms based on her own report. The focal issue I will use as an arbiter at this stage is the analogy between mathematics and science, which will be indeed the backbone of the entire subsequent discussions. The analogy between perception and mathematical intuition found in Gödel is a very unique view that cannot be simply explained away, though it has always been suspected to be too mysterious and dubious. Also, it is important to realize that Quine's indispensability argument, which has been believed to be the most typical argument in favor of mathematical realism, is basically based on the analogy between mathematics and science.

In Sect. 3, I shall briefly discuss Maddy's position after she gave up realism and began to champion mathematical naturalism, primarily based *Naturalism in Mathematics*. As was pointed out above, it is not possible to separate the problem of understanding Maddy's new position from the problem of capturing what has changed in her position. Thus, our project is, in other words, (1) to uncover how Maddy's strategy to meet Benacerraf's challenge has changed, (2) to clarify exactly what aspects of Gödel and Quine were adopted or given up in connection with the analogy between science and mathematics. What, if any, is newly introduced by Maddy? What is implied by introducing these new elements? How exactly are we to understand Maddy's belief change? In executing this project, Maddy's criticism of the indispensability argument is evidently at the center of the debate. Also, it is rather easy to anticipate that, as she perceives it as the most difficult problem for mathematical naturalism, the boundary problem will turn out to be the crucial issue. Ultimately, I intend to show that the problem of the analogy between science and mathematics can supply the better perspective to understand Maddy's changing beliefs. Though too brief, I shall also introduce and examine a few recent criticisms against Maddy in connection with the analogy between science and mathematics In Sect. 4, I shall take an overview of what are meant by the changes in Maddy's position, and present several reasons why I believe Maddy's intellectual journey to be ontological regress.

2 Maddy as Realist

2.1 Maddy's Strategy as a Realist to Meet Benacerraf's Challenge

Until publishing her *Naturalism in Mathematics* (1997), in which she changed her position, Maddy was a long-time typical realist in mathematics by representing what she called "set-theoretic realism". This could be more appropriately characterized as a version of Aristotelianism, though she herself understood it as a revisionistic Platonism. Her work demonstrates how she interacted closely with some of the distinguished practicing mathematicians, thereby showing her exceptionally deep understanding of the recent trend and reality of contemporary mathematics. Above all, the persuasiveness of her arguments was stemming largely from the fact that she had expert knowledge in the most recent advances in set theory.

Benacerraf's epistemological and ontological challenges against the mathematical Platonism around 1970 provide us with an effective watershed for understanding the subsequent developments in philosophy of mathematics. Maddy's case is not the exception to this rule, for her work can be seen as evolved in response to Benacerraf's challenges. Benacerraf's epistemological challenge to Platonism is formulated well as the two horns of the dilemma he launched against platonism (See Benacerraf 1973). The first premise of his argument is the platonist's thesis that mathematical objects are non-spatio-temporal abstract entities. On the other hand, his second premise is the so-called causal theory of knowledge, which was popular and influential at that time, according to which "what makes the belief true must be appropriately causally responsible for that belief" (Maddy 1990, p. 37). For example, in order for me to know that "2 + 2 = 4", the abstract entities 2 and 4 must have causal effects in the generation of my belief. But abstract entities lack causal effectiveness. Thus, if Platonism is correct, we cannot have any mathematical knowledge. On the other hand, if we do have mathematical knowledge, Platonism must be wrong. In other words, we should postulate a special cognitive ability never heard of, or deny that mathematics is about abstract entities such as numbers, sets, or functions. Since neither of the alternatives is satisfactory, now the problem is how to find a way out from this dilemmatic situation. On the other hand, Benacerraf's ontological challenge refers to the problem of so-called multiple reducibility (See Benacerraf 1965). If mathematics describe the domain of abstract entities, as platonists claim, which, among function, space, group, ring, field, could be the populations of the domain? Following the principle of economy, we should not increase the number of entities unnecessarily. According to modern axiomatic set theory, however, since all mathematical objects mentioned above are reduced to sets, the set-theoretic hierarchy can be called the ontology of entire mathematics. But the problem lies in that, in the reduction of arithmetic to set theory, there are multiple ways of doing so, such as von Neumann's method or Zermelo's method. For, depending on which method we adopt, entirely different consequences would follow.

Maddy's strategy to meet Benacerraf's epistemological and ontological challenges is very well known. First, Maddy's strategy to meet Bencerraf's epistemological challenge is to bring down mathematical objects to the spatio-temporal world so that the can contact with our five sense organs. In a word, we can perceive mathematical objects such as sets (Maddy 1990, p. 38f; For further discussion of this issue, see also Levine 2005; MacCallum 2000; Maddy 1980, 1984a, b). For example, suppose that there is a basket that contains three eggs. Maddy claims that then we perceive the set of three eggs. This claim consists of the conjunction of (1) the claim that there are eggs in front of us and (2) the claim that we perceive the set. The importance of the claim (1) can be found in that it clearly departs from the traditional Platonist position, which has denied that sets are located in space and time. Maddy's thesis is that the set of eggs exists exactly where their elements exist. Thus, sets can be located without any difficulty in space and time. As for the more controversial thesis (2), Maddy presents a step-wise argumentation probably in order to block criticisms against it. First, she claims that the numerical belief that there are three eggs in the basket is perceptive. As arguments for it, she points out that we have empirical evidence that such beliefs related to small numbers are non-inferential, and that those beliefs interact non-demonstratively with other clear perceptual beliefs gotten in the situation. Now the problem is why the perceptual experience is about an experience about sets. This is a problem asking what the bearer of the numerical properties are, and Maddy counts it as the problem as to which among the entirety of the physical objects, cconcepts, and sets, the appropriate one to handle the role of the fundamental mathematical entity is. Of course, she opts for the sets. On the other hand, Maddy's strategy to meet Benacerraf's ontological challenge is to treat numbers as properties of sets (Maddy 1990, p. 72 f). However, it seems obvious that even if we treat numbers as properties of sets, Benacerraf's problem would appear again. At this stage, Maddy claims that between the properties that are individuated by the sameness of meaning and the sets individuated by the elements, there is a category of properties individuated by law-like coextensiveness. Accordingly, the predicate 'equinumerous to $\{\emptyset, \{\emptyset\}, \{\emptyset, \{\emptyset\}\}\}$' and the predicate 'equinumerous to $\{\emptyset, \{\emptyset\}, \{\{\emptyset\}\}\}$' are not synonyms, but not accidentally co-extensive, in other words (because their coextensiveness is provable from the axioms of set theory), express the same law-like coextensive scientific property. Since, when understood that way, von-Neumann style numbers and Zermelo style numbers are in fact identical, as a consequence, Maddy claims that according to property theory Benacerraf's dilemma is resolved.

2.2 The Analogy Between Science and Mathematics

I am not particularly interested in plunging into the controversies by assessing the merits and demerits of Maddy's unique position. For I believe that much more important is to understand how, i.e., on what ground and by what process of thought she arrived at such a position. From that perspective, what is notable is that Maddy's strategy to meet Benacerraf's challenge is ultimately inspired by the two Platonist positions in the mid 20th century philosophy of mathematics, i.e., Gödel's Platonism and Quine's Platonism. In fact, Maddy explicitly claims that her set-theoretic realism is a syncretic Platonism based on both. So, it is necessary to check from each what Maddy takes and deserts by paying due attention to both similarities and differences between the two Platonist positions. Above all, it is important to emphasize that both contain suggestive common content that can be subsumed under the general rubric of "the analogy between science and mathematics". In the case of Gödel, the axioms of set theory are so evident enough that mathematicians are bound to accept them. According to him, mathematicians have a special ability of mathematical intuition that allows them to get access to the mathematical truths. Further, he expressed the view that the role of mathematical intuition are rather similar to the role sense perception plays in natural science. On the other hand, in Quine's case, the way of defending mathematical Platonism was very similar to the way of defending common sense realism and scientific realism. For, if an ontology consisting of physical objects and unobservable entities is advocated because it is part of our best theory, mathematics, that not only simplifies physics, and without which we cannot construct physics at all, must be advocated. To be sure, such a thought has been called Quine-Putnam indispensability argument. Then, how did Maddy understand the difference between Gödel's Platonism and Quine's Platonism, and what kind of hints did she get from each in arriving at her position?

Maddy summarizes very clearly how her syncretic platonism was indebted to Gödel and Quine in the concluding part of Maddy (1990) as follows:

> From Quine/Putnamism, it takes the indispensability arguments as supports for the (approximate) truth of classical mathematics. From Gödel, it takes the two-tiered analysis of mathematical justification." (Maddy 1990, p. 178)

Here the two-tiered analysis means the view that the simpler concepts and axioms are internally justified, while the more theoretical hypotheses are externally justified by their consequence.

In the introductory part of Maddy (1990), we once again read the similar passage that summarizes what she was indebted to Quine and Gödel:

> From Quine/Putnam, this compromise takes the centrality of the indispensability arguments; from Gödel, it takes the recognition of purely mathematical forms of evidence and the responsibility for explaining them. (Maddy 1990, p. 35)

Now, let us examine Maddy's theses by using the clues in these paragraphs.

2.2.1   Quine's Platonism and Naturalistic Epistemology: Indispensability Argument

What does it mean that Maddy takes from Quine/Putnamism indispensability arguments as supports for the (approximate) truth of classical mathematics? Exactly what does it mean to take the centrality of the indispensability arguments? Exactly what roles do the indispensability arguments taken in that way play in Maddy (1990)? What exactly did Maddy understand as the indispensability arguments in Maddy (1990)? Since these questions are quite expectable routine ones, we might run the risk of bypassing them by simply assuming that Maddy must have ready answers to them. When we actually raise these questions, however, we find surprisingly that it is hard find any clear answer to them in Maddy (1990).

Maddy presents first the basic idea of the so-called Quine-Putnam indispensability argument. Then, she merely reports some texts from Quine and Putnam's writings, which have been used as supporting evidences without any serious interest to formulate the argument rigorously. According to Maddy, in naturalistic approach, we make judgments about what kind of entities are there by seeing what kind of entities we need in order to produce the most effective theory about the world. Nominalists have no qualms to accepting the ordinary physical objects and the theoretical entities in physics. But the problem lies in how to treat mathematical entities. Maddy is pretty sure that what Quine called "double standard", according to which "question of mathematical existence are linguistic and conventional and questions of physical existence are scientific and real" failed (Maddy 1990, p. 29).

Carnap's various attempts to separate mathematics from natural science culminated in the analytic/synthetic distinction. But, as Quine severely criticized, this distinction turned out to be total failure. Without this distinction, Carnap's anti-Platonist version of logicism was bound to fail (Maddy 1990, pp. 27–28).

It seems that Maddy wants to say that exactly at this stage Quine suggests the idea of the indispensability argument. Since it is not easy to pin down exactly where Quine presents the most rigorously formulated version of the indispensability argument, it is highly revealing that in this context Maddy quotes the following text from Quine:

> A platonistic ontology ... is, from the point of view of a strictly physicalistic conceptual scheme, as much a myth as that physicalistic conceptual scheme itself is for phenomenalism. This higher myth is a good and useful one, in turn, in so far as it simplifies our account of physics. Since mathematics is an integral part of this higher myth, the utility of this myth for physical science is evident enough. (Quine 1948, 18; Maddy 1990, p. 29)

Then, Maddy makes extremely interesting comments on Putnam's views that "mathematics and physics are integrated in such a way that it is not possible to be realist with respect to physical theory and a nominalist with respect to mathematical theory" (Putnam 1975, 74) and that "mathematics and physics are integrated in such a way that it is not possible to be a realist with respect to physical theory and nominalist with respect to mathematical theory" (Putnam 1971, §5 and §7). According to her, "Putnam takes the same thinking somewhat further, emphasizing not only that mathematics simplifies physics, but that physics can't even be formulated without mathematics" (Maddy 1990, p. 29). Also, she is careful enough to quote Putnam's discussion in Putnam (1971, p. 347), which has been counted as the representative text for the Quine/Putnam indispensability argument. There Putnam reminded us of the following facts. According to him, we need to accept the argument, since mathematical entities are indispensable to science. But "this commits us to accepting the existence of the mathematical entities in question". Indeed, Quine already emphasized "both the indispensability of [talk about] mathematical entities and the intellectual dishonesty of denying the existence of what one daily presuppose" (Putnam 1971, 347; Maddy 1990, pp. 29–30).

It is notable that in the subsequent discussion Maddy made the same points as will be made later in Maddy (1997) on Quine. First, Maddy highly evaluated indispensability argument in that, unlike the traditional Platonism, it shows some revolutionary aspects as the following:

> But, if our knowledge of mathematical entities is justified by the role it plays in our empirically supported scientific theory, that knowledge can hardly be classified as a priori" (Maddy 1990, p. 30)

Then, she criticized it for its disagreements with the mathematical practice. First, "unapplied mathematics is completely without justification on this model". Secondly, in its characterization of scientific work mathematics appears only on purely theoretical level so that it "leaves unaccounted for precisely the obviousness of elementary mathematics". In Charles Parsons's phrase, Quine/Putnamism 'leaves unaccounted for precisely the obviousness of elementary mathematics' (Maddy 1990, pp. 30–31. Maddy is borrowing the phrase from Charles Parsons).

We suggested above that it is not clear what it means to say that the syncretic position in Maddy (1990) adopts indispensability argument "as supporting the approximate truth of classical mathematics". Nor is it clear to say that it adopts the centrality of indispensability argument. Now, since we come to know that Maddy is clearly aware of the problem of the status of pure mathematics and the problem of the evidence of elementary mathematics as the fundamental limitations of Quine/ Putnam indispensability argument, we can understand a bit better the meaning of her indebtedness to Quine before she sought the solution from Gödel. For Maddy adds that "successful applications of mathematics give us reason to believe that mathematics is a science, that much of it at least approximates truth. Thus successful applications justify, in a general way, the practice of mathematics" (Maddy 1990, p. 34).

2.2.2 Gödel's Platonism

I claimed above that indispensability argument is based the the analogy between science and mathematics. Interestingly, the contexts in which Maddy discusses the analogy are found mostly the contexts in which she discusses Gödel. In other words, she appears to understand that both Quine/Putnam indispensability argument and Quinean Platonism are irrelevant to the theme of the analogy between science and mathematics. In contrast to this, when she discusses Gödel, Maddy explicitly highlights the analogy between science and mathematics. She carefully distinguishes between the dimensions at which the analogy matters. She even describes the degree of strength of analogy. Here, of course I have in mind not only that Maddy appeals to the analogy between science and mathematics in both the dimension of intrinsic and extrinsic evidence, but also that she describes Gödel's analogy as the strong analogy at the latter dimension. Such an impression is fortified by the fact that she tends to introduce Gödel's analogy immediately after pointing out the problems and limitations of Platonism of Quine/Putnam type. For example, immediately after pointing out that Quine/Putnamism leaves unexplained the evidentness of elementary mathematics, she claims that contrastingly Gödelian Platonism is guided by the real mathematical experience. According to her, the most elementary axioms of set theory are evident to Gödel, and he explained the reason by postulating the special ability of mathematical intuition, which plays a role similar to what sense perception does in physical sciences (Maddy 1990, p. 31: See Gödel 1947/64, Godel, 1990).

According to Maddy, such an analogy between science and mathematics, on which Gödel's Platonism is based, can be in fact traced back to Russell (Maddy 1990, p. 76; Gödel 1944; Russell 1906, 1907, 1919). And Maddy claims that Gödel's analogy contains both ontological and epistemological aspects. As an example of the former, she invokes Gödel's claim that "[Sets] may ... also be conceived as real objects ... existing independently of our definitions and constructions" (Gödel 1944, 456), and, as an example of the latter, she cites Gdöel's another claim that "The analogy between mathematics and a natural science ... compares the axioms of logic and mathematics with the laws of nature and logical evidence with sense perception ..." (Gödel 1944, 449).

Maddy suggests that her task in *Realism in Mathematics* was to replace the Gödelian mathematical intuition, which has been believed to be too mysterious, by bringing down mathematical objects to the spatio-temporal world so that they can be contacted by our five senses. At this stage, Maddy again tries to support the claim that, instead of appealing to mathematical intuition, we can directly perceive mathematical objects in space and time. Forthis, she relies on Donald Hebb's neurophysiological theory (Maddy 1990, p. 55f).

To be sure, such a move is a significant revision of the traditional platonism, as Maddy is clearly aware of, and it is not easy to assess the pay-off.

Maddy's indebtedness to Gödel in connection with the analogy between science and mathematics does not stop there. For, she agrees with Gödel about some important issues: that not all axioms can be justified by appealing to intuition, or that the analogy between science and mathematics should be expanded one step further to the level of scientific hypotheses (Maddy 1990, p. 77). As textual evidence for that thought, she cites from Gödel the following passages:

> the axioms need not necessarily be evident in themselves, but rather their justification lies (exactly as in physics) in the fact that they make it possible for these 'sense perceptions' to be deduced ...(Gödel 1944, 449)

> besides mathematical intuition, there exists another (though only probable) criterion of the truth of mathematical axioms, namely their fruitfulness in mathematics and, one may add, possibly also in physics." (Gödel 1947/64, 485)

2.3   Aristotelian Aspects of Maddy's Realism

We saw above wherein lies Maddy's indebtedness to Quine and Gödel. It seems obvious that she had to go through tortuous paths in distancing herself from them. I am quite convinced that it would be a worthwhile project in itself to reconstruct logically the process of Maddy's changing beliefs. Further, it could be a necessary prerequisite to examine critically the validity of her ultimate position. At this stage, I take a clue from the fact that her syncretic Platonism is in fact more akin to Aristotelianism:

> Some will note that, strictly speaking, this view is more Aristotelian than Platonistic. They are right, in the sense that Aristotle's forms depend on physical instantiations, while Plato's are transcendent. I retain the term "platonism' here, not for its allusion to Plato, but because it has become standard in the philosophy of mathematics for any position that includes the objective existence of mathematical entities. (Maddy 1990, p. 158)

If so, it seems more natural for Maddy to have reflected upon her Aristotelian aspects more deeply and elaborated it. However, she claimed that in case we are to meet Benacerraf's challenge in accordance with her position, the only problem left is to describe and explicate the rationality of thought at the theoretical level. Then, in her later work, she has pursued an entirely different line of thought.

3   Maddy's Mathematical Naturalism: Beyond Quine and Gödel

3.1   What Is Gone, What Is Newly Introduced, and What Is Revised

Maddy's new orientation is clear from the title of Maddy (1997), i.e., *Naturalism in Mathematics*. But what does it mean that mathematical realist Maddy changed into mathematical naturalist? One of the necessary reasons to raise this question is that Maddy herself seem to emphasize both continuity and discontinuity of her two books. As the evidence for discontinuity, we may cite the structure of Maddy (1997) itself, i.e., the fact that Maddy compares the solution from mathematical realist camp and that from mathematical naturalist camp to the problems of the independent axioms in axiomatic set theory, and argues for the superiority of the latter As for the continuity, we may point out that apparently the biggest problem in Maddy (1997) is the problem of the independent axioms in modern axiomatic set theory, and that was the very problem identified by Maddy (1990) as the most important open problems of mathematics. Furthermore, she not only has paid unusually careful attention to the practice of mathematicians (including Gödel) in axiomatic set theory, but also has been sympathetic with Quine's naturalized epistemology. All this indicates that from the beginning she has a strong tendency toward naturalism. If so, in what respects did she expand the naturalism adding new elements? What does that line of thought, which emphasizes naturalism so much, imply to her ontological position? Does it lie on the right track?

When we raise questions like these by focusing on the new aspects of Maddy's position, the following point seems notable. One consequence of Quine's indispensability argument is that only applied mathematics is justified, and pure mathematics is treated as an uninterpreted system. It seems that Maddy has some qualms with it. In fact, she questions the validity of indispensability argument itself, by discussing the limit of applying indispensability argument, and by casting a doubt on the assumption that the interaction of mathematics and science has implications to ontology of mathematics. Such a doubt is led to an observation that mathematicians and scientists themselves are neither interested in the application of mathematics in science, nor are they work as indispensability argument presumes. Then Maddy's mathematical naturalism comes to the front. Even though mathematics is central in natural science, the method of mathematicians are entirely different from the method of natural scientists. Maddy wants to emphasize that, other than proof and the axiomatic method, there is no tribunal for the mathematicial method.

On the other hand, we may focus on what is gone in Maddy's changed position, i.e., her abandonment of realism. Her abandonment of realism was rather shocking to many, and, if we understand the change as change from realism to anti-realism, the shock could be more dramatically amplified. In fact, there were scholars who understood what was going on precisely that way (See Chihara 2004; Also, see Chihara 1982, 1990, 1998), and Maddy had to struggle with them in order to correct some such misunderstanding. The problem lies in that, even if her changing belief was not from realism to anti-realism, evidently it was an abandonment of her previous realist position. The far-reaching influence of such an abandonment of realism can be felt from the fact that the discussion of Benacerraf's challenge to Platonism, which was a central issue in Maddy (1990), is rarely found in Maddy (1997). Probably, in Maddy (1997), such a discussion would be irrelevant to the thrust of the main argument. Then, why and how did Maddy abandon her previous realist position?

Colyvan seems to provide us with a clear answer to this question. It is because Maddy came to criticize indispensability argument she previously adopted whole-heartedly. This answer seems to be based on an implicit reasoning like the following. By criticizing indispensability argument, which was believed to be the only good argument without begging the question in favor of realism, consequently there is no good argument left for realism. Thus, it is a necessary consequence that Maddy "has renounced realism she so enthusiastically argued for" (Colyvan 2001, p. 91). Though there are some aspects in this reasoning hard to accept, let us find some clues from Colyvan's attractive answer.

In view of its scope and the complexity of its content, Colyvan's slim book *The Indispensability of Mathematics* (Colyvan 2001) is potentially controversial. Insofar as the so-called Quine-Putnam indispensability argument is one of the most central issues in contemporary philosophy of mathematics, however, this book is destined to be of focal interest. Colyvan starts Chapter 5 with the following suggestive observations:

> In a series of recent essays,* one-time mathematical realist Penelope Maddy has presented some serious objections to the indispensability argument. Indeed, so serious are these objections, that she renounced the realism she so enthusiastically argued for in Maddy (1990)" (Colyvan 2001, p. 91) (*See Maddy 1992, 1995, 1996c, 1998b: This is Colyvan's footnote.)

The reason why the first sentence of the quoted paragraph is suggestive can be found in that it treats the position of Maddy (1990) toward the indispensability argument simply as an objection. On the other hand, the reason why the second sentence is suggestive can be found from the fact that it provides us with a clear and promising answer to the question as to why Maddy changed her position from Maddy (1990) to Maddy (1997). But are these two sentences true?

After examination, though reluctantly, I have to concede that they are. But, even if Colyvan's two sentences are true, didn't he fail to give enough ground for their truth.[1] To be sure, Maddy (1997) criticized indispensability argument. Colyvan summarized her criticisms as three arguments, and then tried to criticized them. Of course, Maddy (1997) abandoned the realist position she sustained until Maddy (1990). Nevertheless, it necessary to give up the realist position? Shouldn't we find out the crucial reason for the abandonment of realism from somewhere else?

No doubt, it must be a significant clue that naturalism comes to the foreground. However, naturalistic orientation was already there in Maddy (1990). This can be witnessed by her even earlier writings. As a consequence, we may conjecture that Maddy's meta-philosophical reflections around naturalism would be more important in understanding Maddy's belied change. From that point of view, we need to examine carefully Maddy's arguments in her criticism of Quine's indispensability argument. By now, a working hypothesis that Maddy arrived at the position in Maddy (1997) by subtly distorting Gödel's position emerges as an attractive possibility. However, the most important clue we should rely on in the process of tracking the true reason for Maddy's belief change seems to be the analogy of science and mathematics, which seems to have contributed the big framework for thought on a macroscopic dimension.

We saw above in Sect. 2 how Maddy was indebted to in her mathematical realist years to Quine and Gödel within the framework of the analogy between science and mathematics. If so, it seems natural to expect that this framework would be instrumental in understanding her new position after abandoning mathematical realism. Did Maddy attempt to build her new position by going beyond Quine and

---

[1]This doubt stemming from a cursory reading of chapter five of Colyvan's book is now becoming more and more serious. According to him, the chapter largely depend on an independent paper, i.e., Colyvan (1998), and the review, i.e., Colyvan (1999). Except for the fact that the introductory part of the chapter five was rewritten and the last part of the conclusion was added, it is simply reprint of Colyvan (1998). There seems to be no newly added content based on his review of Maddy's book. If so, the argument for the two troublesome sentences may not have been fully presented.

Gödel without leaving the framework of the analogy between science and mathematics? Or, was she able to arrive at her new position clearly distinct from Quine and Gödel only by abandoning the analogy between science and mathematics itself? We need to be alert to identify what are gone without any trace, what appeared newly, and what were revised, in the process of transformation from Maddy as mathematical realist to Maddy as mathematical naturalist. Also, we may rely on a conjecture that in the process Maddy's changing beliefs regarding the analogy between science and mathematics. With all this in mind, let us examine the two salient issues, i.e., Maddy's criticism of indispensability argument, and the so-called boundary problem.

3.2   The Analogy Between Science and Mathematics Revisited

3.2.1   Maddy's Criticism of Indispensability Argument

Though Quine/Putnam indispensability argument has been widely discussed, it was Maddy (1992, 1997) who first formulated the indispensability argument rather rigorously by enumerating and characterizing each and every premise. We can notice a clear difference between Maddy (1992) and Maddy(1997) on the one hand and Maddy (1990) on the other in her ways of treating this argument. As we pointed out above, in Maddy (1990), she merely presented the basic idea of the argument and gave some textual evidence from Quine and Putnam. There was serious attempt to formulate the argument in rigorous fashion. On the other hand, now she attempts a rigorous formulation of the argument, by enumerating all the necessary premises of the argument, and examining the character of each of them.

According to her, it can be represented as follows:

Premise 1: Our best scientific theory of the world makes indispensable use of mathematical things. (fact)

Premise 2: To draw a testable consequence from our theory requires the use of various far-flung parts of the theory, including much mathematics, so the confirmation resulting from a successful test adheres not to individual statements but to large bodies of theory. (holism)

Premise 3: Our theory is committed to those things that it says 'there are'. (Quine's criterion of ontological commitment)

Therefore, Our theory is committed to the existence of mathematical things. (Maddy 1997, p. 133)

Colyvan examines Maddy's criticisms of indispensability argument by categorizing them into three: (1) scientific fictions objection, (2) the role of mathematics in science, and (3) the mathematical practice objection. On the other hand, Lee distinguishes first between the objection related to the autonomy of mathematics and the objection relevant to the logical structure of indispensability argument. Then, he subdivides the latter into (1) criticism based on history of science, (2) criticism about the relationship between mathematics and science, and (3) criticism related to the attitudes of practicing scientists toward mathematics (Lee 2005, 119–122). The mathematical practice objection of Colyvan seems to be what Lee calls the objection related to the autonomy of mathematics, and was dealt with in Chapter 7 of Maddy (1997). Colyvan's scientific fictions objection is nothing but Lee's criticism based on history of science, and was discussed in Chapter 6 of Maddy (1997, pp. 135–143). What Colyvan calls the objection about the role of mathematics in science is what Lee calls crticicism about the relationship between mathematics and science, and was covered by Chapter 6 of Maddy (1990, pp. 143–152). As Lee points out, the so-called criticism based on history of science and the criticism related to the attitudes of practicing scientists toward mathematics can be lumped together. Thus, there is room for understanding that both Colyvan and Lee are examining the same arguments in their own ways of discussion. However, according to Maddy, while both scientific fictions objection based on history of science and the objection about the role of mathematics in science are criticisms about the scope of indispensability argument, criticism related to the attitudes of practicing scientists toward mathematics is a criticism against the validity of indispensability argument itself. Thus, it seems to be prudent to categorize objections in accordance with Maddy's initial discussion. The criticism related to the attitudes of practicing scientists toward mathematics is found in Maddy (1997, pp. 154–157).

The so-called scientific fictions objection is adopted from the history of atomism covering the period from the early 19th century when it was first introduced to the early 20th century when it was universally accepted. Let us the follow the summarized presentation of Colyvan, for it is not necessary for my present purpose to examine Maddy's detailed exposition. As Colyvan aptly points out, the puzzle foe the Quinian "is to distinguish between the situation in 1860, when the atom became 'the fundamental unit of chemistry' and that in 1913, when it was accepted as real" (Colyvan 2001, p. 92; Maddy 1994, 394). The point is that of Quineans are right, then scientists around 1860, when atoms turned out to be indispensable in their theory, should have accepted the existence of atoms. However, some leading scientist, including Poincaré, were skeptical about the existence of atoms even until 1904. From this example, Maddy derives the lesson that "scientists do not take the indispensable appearance of an entity in our best theory to warrant the ontological conclusion that it is real" (Maddy 1997, p. 152). So, she claims that "strong uses of indispensability … are in conflict with the practice of science". [Ibid.]

Given the lessons from the case of atomism, Maddy thinks, we have to examine more carefully the details of how mathematics appears and functions in science (Maddy 1997, p. 143). Then, she presents another objection related to the role of mathematics in science, with focusing on the fact that "many of the applications of mathematics occur in the company of assumptions that we know to be literally false" (Maddy 1997, p. 143; Colyvan 2001, pp. 92–93). For example, "we treat a section of the earth's surface as flat, rather than curved, when we compute trajectories" (Maddy 1997, p. 143). And, as a consequence, she even mock the indispensability based on the application of mathematics to science by a rhetorical question:

> On the face of it, an indispensability argument based on such an application of mathematics in science would be laughable: should we believe in the infinite because it plays an indispensable role in our best scientific account of water waves? (Maddy 1997, p. 143)

The objection regarding the attitudes of practicing scientists toward mathematics starts with the concern that science is not done as it would be if it were done as indispensability argument requires, and especially the concern that it is not be done as it would be if it were the measure of mathematical ontology (Maddy 1997, p. 154). Maddy thinks that physicists happily use any mathematics whatsoever, as long as it is useful and effective, without any interest in mathematical existence assumptions, and without any interest in structural assumptions of physics presupposed by mathematics (Maddy 1997, p. 155). Thus, it is Maddy's thought that the structural assumptions of physics underlying the assumptions of mathematical existence and the application of mathematics do not have epistemic standard of the equal level as ordinary physical assumptions, and their role in successful theory would lack the conformational power (Maddy 1997, p. 156).

As was pointed out, what Colyvan calls the mathematical Practice objection appeared already in Maddy (1990), and originated from the dissatisfaction that Quine/Putnam indispensability argument does not grant a legitimate status to pure mathematics. This task done in Chapter 7 of Maddy (1997) is to show that, if in more subtle way, the same problem still remains even if we revise the original Quine/Putnam indispensability argument. By selecting axiomatic set theory as the obvious example of pure mathematics, Maddy presents an argument this way. If it turns out that there is a literal application, the problem of the independent axioms of axiomatic set theory would have solutions. If it turns out that there is no literal application, i.e., if it turns out that if all applications of continuum mathematics are merely idealizations, realist argument would collapse, and there would be no fact of the matter (to distinguish between truth and falsehood. If this thought is right, then we would expect that set theorists, in evaluating the application of continuum mathematics to natural science, would be most interested in the recent trend in quantum field theory. But the problem is that set theorists have no interest at all about such problems. So, Maddy conjectures that no matter how these issues fare in natural science, the practice of set theory, and the method of set theorists dealing with the problem of independent axioms would not be affected at all (Maddy 1997, pp. 158–159).

Now it becomes important to capture the structure and character of maddy's four objections to the indispensability argument. Prior to it, however, we need to discuss briefly two problems regarding the indispensability argument as her target. First, is it all right to understand the indispensability argument, as Maddy formulates? Secondly, how are we to understand the basic structure of the indispensability argument? In fact, these two problems are intertwined.

I believe that it is necessary to raise the following two questions, which are closely related, before plunging into debates over the indispensability argument: (1) Is it all right to accept Maddy's formulation as capturing the essence of the indispensability argument? (2) How are we to understand the basic structure of the indispensability argument?[2]

What I mean by the first question is this. In Maddy's formulation, Quine's holism is one of the necessary premises. But I do not understand why that should be the case. Probably, I may invoke the fact that Quine's holism did not play any significant role in her earlier treatment of the indispensability argument (For example, in Maddy 1990). Nor do I find any evidence that Quine or Putnam counted holism as an essential part of the indispensability argument. It is not difficult to surmise the reason why no objection has been made against counting holism as a necessary element in the indispensability argument. For, holism is one of the famous theses in Quine's thought, and its content is apparently relevant to the context in which the indispensability argument is invoked.

Maddy's desire to interpret the indispensability argument as aiming at a deductively valid argument seems to be responsible for her decision to include holism as a necessary part of the true indispensability argument. I am not even sure whether the difference between including and excluding holism would determine the validity of the indispensability argument. In fact, I am not interested in some such problems at all. For, I believe that the idea itself to understand the indispensability argument as aiming at a deductively valid one must be the result of misunderstanding its basic structure and character.[3] Indeed I believe that the indispensability argument is basically conceived as an analogical inference based on the analogy of science and mathematics.

If so, we may even formulate such an analogical inference by adding two new premises instead of holism: (1) Our scientific theory says that the theoretical entities included in it do exist; (2) Our scientific theory has indispensable uses of the theoretical entities included in it.

Premise 1: Our best scientific theory of the world makes indispensable use of mathematical things. (fact)

Premise 2: Our scientific theory has indispensable uses of the theoretical entities included in it.

Premise 3: Our scientific theory says that the theoretical entities included in it do exist.

Premise 4: Our theory is committed to those things that it says 'there are'. (Quine's criterion of ontological commitment)

Therefore, Our theory is committed to the existence of mathematical things.

---

[2]There is a huge literature on the indispensability argument. For those that were influential in the earlier period, see Burgess nd Rosen (1997), Burgess (2004), Balaguer (1994, 1996), Decock (2003), Colyvan (1998), Leng (2002), Peressini (1997, 2003), Resnik (1995), Risikin (1994), Sober (1993), Tieszen (1994).

[3]It is not difficult to find examples where the indispensability argument is considered as aiming at a deductively valid argument. By citing Maddy (1992) and Sober (1993), Peressini points out that scholars began to discuss the validity of the indispensability argument rather than following Field's strategy of denying the thesis of the indispensability of mathematics (Peressini 1997, 211). Perhaps, the best example can be found in Pincock. For, he attempts to argue "Colyvan's argument does not imply its conclusion" (Pincock 2004, p. 62).

Alternatively, we may formulate the analogical inference in the following fashion:

Premise 1: Mathematical objects are similar to the theoretical entities in science in that they have indispensable uses in our best scientific theory.

Premise 2: Our scientific theory says that the theoretical entities included in it do exist.

Premise 3: Our theory is committed to those things that it says 'there are'.

Therefore, Our theory is committed to the existence of mathematical things.

It even seems to me that these new premises were so obvious that they were omitted by Quine/Putnam indispensability argument.

What consequences would follow if I am on the right track? In that case, those critics who attempt to criticize the indispensability argument (as formulated by Maddy) by attacking Quine's holism must be committing a fallacy of straw man. Also, all attempts to defend or challenge the deductive validity of the indispensability argument should be rejected as pointless.

Now we are in a position to evaluate Maddy's objections to the indispensability argument. If the indispensability argument is meant to be an analogical inference, our evaluation of her objections to the indispensability argument must depend on (1) whether Maddy herself understands it as an analogical inference, and (2) whether she assumes that her opponents understand it as an analogical inference.

We have reason to answer positively to the first question. For, unless we assume such an answer in at least one case, Maddy's strategy itself would be nullified. According to Maddy, mathematical existence assumptions in science are not treated on a par with ordinary physical assumptions. Further, she claims that this epistemic disanalogy undermines the groundwork of the original Quinean argument (Maddy 1997, p. 156). It must be meaningless to point out such a disanalogy in this context, unless she considers the indispensability argument as basically an analogical inference. Apparently, Maddy appears to be rather inconsistent in treating the analogical character of the indispensability argument.

There are more difficulties in answering the second question. For, this problem is connected to another problem of understanding Maddy's motivation to include Quine's holism as a part of the indispensability argument. If Maddy herself counts it as an analogical inference, but she assumes that her opponents don't, then we would have to criticize Maddy for her abusing the weakness of her opponents in her scientific fiction objection and her objection related to the role of mathematics in science (cf. Colyvan 2001). For, these two objections are merely pointing out some trivial facts, having no potential as criticisms of an analogical inference. In case both Maddy and her adversaries assume it as an analogical inference, we would have to examine these two objections much more carefully. And it would be meaningful to distinguish between parts of a theory that are true and that are merely useful (Cf. Maddy 1992, 281; Colyvan 2001, p. 92). However, they would still be powerless as criticisms against an analogical inference.

All this indicates that the indispensability argument is basically an analogical inference, and that both defenders and challengers should have fully recognized the analogical character of the indispensability argument. This conclusion seems further supported by Maddy's mathematical practice objection (cf. Colyvan 2001). In order for the strategy to reveal the invalidity of the indispensability argument by *reductio ad absurdum* to get off the ground, the absurdity of imagining an axiomatic set theorist who pays utmost attention to the recent developments of quantum field theory should be evident. In other words, Maddy is begging the question by smuggling the disanalogy of mathematics and science as an implicit premise in her reasoning. If a disanalogy counts as an effective criticism, the target of the criticism must be an analogical inference.

3.2.2   The Boundary Problem

Maddy calls the problem as to "whether or not philosophical debates on the ontological status of mathematical objects are integral parts of mathematical practice, whether or not they are 'continuous with mathematics', whether or not they are external, extra-mathematical "the boundary problem". And, according to her, it is the most difficult problem for mathematical naturalists (Maddy 1997, p. 188).

In order to understand why this matters so much, we need to point out that this does not merely question the boundary between mathematics and philosophy. It does also question the boundary between mathematics and science, and the boundary between science and philosophy. It seems obvious that Maddy has all these in mind, for only on the basis of what she discussed the other two problems to a certain extent she was able to announce the problem of the boundary between mathematics and philosophy as the most difficult problem for mathematical naturalists. In other words, Maddy already eliminated the first philosophy via her discussion of Quine's position, and fixed the boundary between mathematics and science again via her discussion of Quine. We should understand that only against that background she is now dealing with the other remaining problems. If so, we need to briefly scheme what Maddy is presupposing here, i.e., he discussion of Quine's position about the boundary issue.

Maddy confesses her indebtedness to Quine almost everywhere. According to Quine, naturalism is "the recognition that it is within science itself, and not in some prior philosophy, that reality is to be identified and described" (Quine 1981, 21; Maddy 1997, p. 180). Maddy also adopts wholeheartedly Quine's understanding of naturalism as follows:

> abandonment of the goal of a first philosophy. It sees natural science as an inquiry into reality, fallible and corrigible but not answerable to any supra-scientific tribunal, and not in need of any justification beyond observation and the hypothetico-deductive method". (Quine 1975, 72; Maddy 1997, p. 180)

It is in connection with the problem of the relation between mathematics and and science where Maddy begins to distance herself from Quine. According to Maddy, her mathematical naturalist "begins within natural science". But it becomes immediately obvious that "mathematics is central to our scientific study of the world and that the methods of mathematics differ markedly from those of natural science" (Maddy 1997, p. 183). But "Quinean naturalist persists in subordinating mathematics to science, on identifying the proper methods of mathematics with the methods of science" (Maddy 1997, p. 184).

At this stage, Maddy even claims that "to judge mathematical methods from any vantage-point outside mathematics" is "to run counter to the fundamental spirit that underlies all naturalism". Here, the so-called "the fundamental spirit" is according to her "the conviction that a successful enterprise … should be understood and evaluated on its own terms". So, Maddy proposes to pay the same respect to mathematical practice as Quine paid to science. Borrowing Quine's celebrated expression, she announces that "mathematics is not answerable to any extra-mathematical tribunal and not need of any justification beyond proof and the axiomatic method" (Maddy 1997, p. 184). And she elaborates that, as Quine treats science as independent of the first philosophy, she treats mathematics as independent of both the first philosophy and natural science (including the naturalized philosophy continuous with science).

In understanding maddy's problem of boundaries between mathematics and philosophy, I think, we can have a well-rounded perspective once again by focusing on the analogy between science and mathematics. For example, we may think this way. Quine wanted to derive an analogical inference like indispensability argument not merely satisfying with the analogy between science and mathematics at level of metaphor or suggestion. As a consequence, the borderline between science and mathematics was blurred, and thoughts like Quine's gradualism or conformational holism were resulted. It seems that Maddy, by criticizing Quine/Putnam indispensability argument, and by including conformational holism as one of its premises, intended to give the appearance that she treated Quine's thought related to the analogy between science and mathematics in wholesale fashion. If we accept indispensability argument as an argument in favor of mathematical realism, it becomes difficult to explain the status of pure mathematics. Let us not forget the case of Quine, who treated pure mathematics as recreation.

On the other hand, the situation is different in Gödel. In Gödel, it is difficult to find instances in which he derives certain particular inference, such as indispensability argument, based on the analogy between science and mathematics. For that very reason, it could more difficult to criticize Gödel's thought related to the analogy between science and mathematics.

If mathematics has secured the independent status both from philosophy and science, for what other remaining problems did Maddy have to discuss the boundary between mathematics and philosophy once more? The answer seems evident. That was due to Gödel, for he was "an unabashed practitioner of first philosophy" (Maddy 1997, 183). In other words, the problem Massy is facing in her discussion of the boundary between mathematics and philosophy is nothing but how to differentiate her position from that of Gödel. Given my discussion above, that problem is exactly the problem of how to how to pay back for her indebtedness in Maddy (1990) in connection with the analogy between mathematics and science to Gödel.

3.2.3 Gödel as Mathematician and Gödel as Philosopher: The Problem of Independent Axioms and the Analogy Between Science and Mathematics

Needless to say, Gödel was a mathematician. He was a central figure in the history of 20th century axiomatic set theory, which is Maddy's focal interest. The method of Gödel as mathematician in studying continuum hypothesis or the axiom of constructability is peculiar to mathematics. In other words, Gödel's method deserves to be admired by mathematical naturalists. But, as Maddy correctly observes, Gödel was an abashed philosophy, who practiced the first philosophy without any shame. Is it possible at all to sharply distinguish between mathematics and philosophy? Didn't Maddy unjustifiably distinguish between Gödel as mathematician and Gödel as philosopher, paying wholehearted respect to the form, while ignoring or distorting the latter? If my interpretation that indispensability argument is basically an analogical inference based on the analogy between science and mathematics, it would become utterly important to uncover the place, function, and the character of the analogy in Quine's thought. Nevertheless, the problem of the analogy between science and mathematics seems even more important in Gödel. Maddy was not blind to that point, in view of the fact that she discussed the analogy more extensively in the contexts of dealing with Gödel rather than Quine. It seems to me, after all, that is not because the analogy comes to the fore by Gödel's emphasis, but because the firm conviction about the analogy between science and mathematics was embodied in Gödel's work itself. We should appreciate Maddy's perceptiveness for her in-depth exposition in that regard. But, in what sense is the analogy between science and mathematics embodied in Gödel's work itself?

As I pointed out above, Maddy (1997) is centering around the problem of the independent axioms in axiomatic set theory. And, the comparison of the the approach of mathematical realism and the approach of mathematical naturalism newly represented by Maddy is the guiding strategy. Now, what needs to be emphasized is that the former, i.e., the approach of mathematical realism, is nothing but Gödel's approach to the analogy between science and mathematics.

Maddy, in the part 2, Chapter 4 of Maddy (1997), presents the objections of the mathematical realists against the axiom of constructibility ($V = L$). Here, she finds the fundamental reason for the researchers in axiomatic set theory to oppose to this particular candidate for an axiom in that it unjustifiably restricts the notion of arbitrary set of integers. While it requires "every set to be definable in a certain uniform way", such a requirement is, "implausible, undesirable, to be avoid" (Maddy 1997, p. 110; See also Moschovakis 1980, 610).

And, Maddy claims that in this stream of thought mathematical realism plays two separate roles. First, realism wants to emphasize that it is not enough to say that "ZFC + $V = L$ and ZFC + $V \neq L$ are equally acceptable because they are equiconsistent with ZFC", and there is some real problem to be solved. Secondly, according to Maddy, set-theoretic realism inherits from Gödel "the strong sscience/ mathematics analogy. Maddy thinks that this implies that "in mathematics, as in science, there are legitimate forms of evidence beyond the sensory and the intuitive".

What Maddy means by "evidence" here is "evidence based on the consequences of a given hypothesis, or more generally, on the virtues of the type of theory it produces". She calls such an evidence 'extrinsic evidence' (Maddy 1997, 110–111).

Here is Maddy's sketch of the mathematical realists' strategy to oppose to $V = L$ in terms of the analogy between science and mathematics:

> I describe a mathematical maxim with a history similar to that of Mechanism (in (ii)), and I suggest that $V = L$ is suspicious in light of its connections with the maxim (in (iii)). Assuming the science/mathematics analogy, if it is rational to proceed as the natural scientists did, it should be rational for set theorists to hold this attitude towards $V = L$. (Maddy 1997, p. 116)

The mathematical maxim here refers to the so-called definabilism, Maddy in fact describes briefly the rise and the fall of definabilism. As mechanism faced various anomalies and superseded by the concept of field, definabilism also faced a series of anaomalies and ultimately replaced by another maxim, i.e., combinatorialism.

Until so far, we have confirmed that Gödel was a mathematical realist, and that he relied heavily on the analogy between science and mathematics. Further, we noted that these two facts are inseparably interrelated in Gödel's philosophy of mathematics, which placed a central role in his work on solving the problems in axiomatic set theory. In her realist years, Maddy welcomed all this, and tried to develop mathematical realism modeled after it. Then, for what reason did Maddy abandon mathematical realism and come to advocate mathematical naturalism? What does this change mean to the theme of the analogy between science and mathematics? Also, what does this mean to the relationship between Gödel and Maddy?

A clue for all these questions should be found in part 3, Chapters 3 and 4 of Maddy (1997). The outline is somewhat like this. In part 3, Chapter 4, she wants to highlight Gödel, who dealt with the problem of the independent axioms in axiomatic set theory by purely mathematical considerations without depending on the analogy between science and mathematics. By such a work, Maddy intends to establish the strategy and methodology of mathematical naturalist clearly distinguished from that of mathematical realist. In other words, Maddy wants to highlight Gödel as mathematician, contradistinction to Gödel as philosopher, in order to make the former as the model for mathematical naturalists.

It is already obvious that such a project is an arduous task. Above all, Maddy herself frankly concedes that she has "no principled distinction between the philosophical and the mathematical" (Maddy 1997, p. 193). A few lines later, she also writes: "Disappointing as this may be, I think there is no principled distinction to draw between mathematics and philosophy, between mathematics and science". Likewise, she also concedes that "the boundary between science and naturalized philosophy, on the one hand, and first philosophy, on the other, is no doubt vague in places" (Maddy 1997, p. 190). She is well aware of the fact that throughout the history of mathematics debates in mathematics have been mixed up with ontological discussions (Maddy 1997, 187). Above all, in the case of Gödel, it must be extremely difficult, even if not impossible, for Gödel himself self-consciously represented and argued in favor of a philosophical position (mathematical realism), where Maddy would expect only the purely mathematical. Then, how does Maddy work out this difficult task?

Surprisingly enough, she executes this task relatively persuasively. Her strategy is to show that, in dealing with the real problems in axiomatic set theory, Gödel first derived the conclusion on purely mathematical considerations, and only then introduced philosophical speculations. A nice example, in which this strategy is well exemplified, is the text in which Gödel's criticism of Russell's vicious circle principle was discussed. First, Maddy quotes the following from Gödel:

> It is demonstrable that the formalism of classical mathematics does not satisfy the vicious circle principle....since the axioms imply the existence of real numbers definable in this formalism only by reference to all real numbers. Since classical mathematics can be built up on the basis of *Principia* (including the axiom of reducibility), it follows that even *Principia*... does not satisfy the vicious principle. (Gödel 1944, 127; Maddy 1997, p. 174; See also Russell 1906, 1907, 1919, 1973)

She immediately draws our attention to Gödel's conclusion that "I would consider this rather as a proof that the vicious circle principle is false than that classical mathematics is false" [ibid.] Then, Maddy develops the following argument:

> What stands out, for our purposes, is that the argument does *not*run:the VCP is an Anti-realist claim: Realism is correct for reasons x, y, and z; therefore, the VCP is false. Rather, the argument goes straight from mathematical actualities to the falsity of the VCP, without any detour through philosophy. The philosophical theorizing begins only *after* the conclusion against the VCP has been drawn." (Maddy 1997, p. 174)

To be sure, in order for such a strategy to be effective, we should presuppose the distinction between mathematics and philosophy, the extreme difficulty of which was already pointed out. Maddy's maneuver to overcome the difficulty is remarkably subtle. Regarding Gödel's critical discussion of Russell's no-class theory, Maddy is anxious to show that exactly where we might expect Gödel to employ the analogy between science and mathematics, there appeared some purely mathematical considerations. According to her, exactly where we might expect him to criticize no class theory as he criticized phenomenalism, instead of appealing to the analogy between science and mathematics, Gödel criticized it by saying that" the classes ... introduced in this way do not have all the properties required for their use in mathematics" (Gödel 1944, 132; Maddy 1997, p. 173). And Maddy notes that here the existence assumption of sets does not depend on either the mathematical analogue of of sense perception or the rigid analogy between science and mathematics, but rather on the conditions of ordinary mathematical practice.

By assuming all these discussions, Maddy even suggests the irrelevance of philosophical realism to what Gödel was truly interested in. Maddy's quotation from Gödel as a supporting argument seems surprisingly fits well to her purpose:

> However, the question of the objective existence of the objects of mathematical intuition ... is not decisive for the problem under discussion here [i.e. the meaningfulness of the continuum problem]. The mere psychological fact of the existence of an intuition which is sufficiently clear to produce the axioms of set theory and an open series of extensions of

them suffices to give meaning to the question of the truth or falsity of propositions like Cantor's continuum hypothesis. [Gödel, ibid.]

Maddy once again emphasizes that the point that what is crucial is mathematical considerations rather than philosophical ones is crystal clear here. Also, she emphasizes that Gödel's interest lies in specific issues in real mathematics rather than in producing supporting arguments for mathematical realism based on the analogy between science and mathematics.

Maddy is not denying the fact that we can find the analogy between science and mathematics and the philosophy of mathematical realism in Gödel's writings. Rather she is claiming that a second stream of thought, that is different from these and is in tension with them, is found in Gödel. It is that second Gödelian theme Maddy wants to develop into the mathematical naturalism.

However, even if we fully concede that Maddy successfully distinguished between Gödel as mathematician and Gödel as philosopher, which is indeed an arduous task, and even if we appreciate the originality of her attempt to develop her own mathematical naturalism from there, there still remains a serious suspicion as to the possibility that she distort Gödel's positionably. As we saw above, Maddy emphasizes that in Gödel only after a conclusion has been secured there follows philosophical considerations. For example, it was the case where she mentions Gödel's discussion of Russell's principle of vicious circle. However, it is always a vexing problem to determine the priority or dependence relations, and especially so when there are complicated relations among the motivation or purpose and justification of a certain position. In order to highlight this point, I will compare Cassou-Noguès' argument and Maddy's argument, which seem to show the same pattern:

> One may use impredicative definitions, he asserts, if and only if one refuses the constructivist conception of mathematical objects and, so it seems, accepts the Platonistic view. But impredicative definitions are indispensable in classical mathematics. Therefore, in order to save classical mathematics, one has to accept the Platonistic view. (Cassou-Noguès 2005, p. 216)

> I suggest that the motivation for Gödel's Platonism is his desire to maintain the meaningfulness of a mathematical pursuit of the Continuum Hypothesis despite its independence of the standard axioms. It might be thought that he believes in the meaningfulness of CH because he believes in Platonism, but I suggest that the dependency goes the other way around. (Maddy 1996a, p. 69)

At first sight, there seems to be no serious reason to complain against Cassou-Noguès' argument from Maddy's position. For, it would be possible to interpret that it was motivated by classical mathematics, and the philosophical stance was accordingly determined. However, Maddy should be careful in responding to the idea that "we should adopt Platonism in order to save classical mathematics", for, in such a case, it would be impossible to say that an irrelevant philosophical position was supervenient to the conclusion arrived by purely mathematical consideration. Above we emphasized the fact that Gödel's interest lies in some specific issues in real mathematics rather than in the production of supporting arguments for mathematical realism based on the analogy between science and mathematics.However, Cassou-Nogués presents an argument like this in the context of finding in Gödel a supporting argument in favor of Platonism, and her position is contrary to that of Maddy's. In fact, according to Cassou-Nogués, Gödel pursued in various ways to formulate a supporting argument for platonism (Cassou-Nogués 2005, Sects. 3–4). On the other hand, Cassou-Noguès may not have much complaint against Maddy's discussion. For, if we just remember his use of bi-conditional in his reconstructed argument, the direction of the dependence does not seem to bea serious matter, as is in Maddy. For him, it seems sufficient if we can be sure that on the same ground Platonism was needed in order to save the mathematical pursuit of continuum hypothesis. In other words, while we need to praise Maddy for her reminder of the reasonable claim that the mathematical consideration of Gödel as mathematician motivated his philosophical position, I think, we should be skeptical about her attempt to put too much emphasis on the claim so as to consider the formulation of a supporting argument in favor of Platonism in Gödel a subsidiary philosophical project irrelevant to mathematical considerations.

3.2.4 Why Was the Analogy Between Science and Mathematics Given up?

We found above that Maddy (1997) is centering around the problem of the independent axioms in axiomatic set theory. Maddy's guiding strategy turns out to be comparing the mathematical realist's and Maddy's mathematical naturalist's approaches to the problem. We also noted that mathematical realist's approach is nothing but Gödel's approach to the analogy between science and mathematics. Now the only problem left to Maddy, who wants to initiate the mathematical naturalist program by sharply distinguishing between Gödel as mathematician and Gödel as philosopher, and by exclusively appealing to the former, is how to reject the Gödelian mathematical realist approach to the analogy between science and mathematics. By what kind of discussion does Maddy reject the analogy between science and mathematics? Maddy explains how she has become skeptical about her previous realist position by an autobiographical remarks in Chapter 5 of part 2 of Maddy (1997) as follows:

> To begin with, among the various justifications proposed for $0^{\#}$ and the rest, the most compelling seem to rest on maximizing principles of a sort quite unlike anything that turns up in the practice of natural science: crudely, the scientist posits only those entities without which she cannot account for our observations, while the set theorist posits as many entities as she can, short of inconsistency. In fact, this very contrast seems to lie behind the poor fir between pure Quinean indispensability and set theoretic practice: Quine counsels us to economize, like good natural scientists, and thus to prefer $V = L$, while actual set theorists reject $V = L$ for its miserliness. This raises questions about the viability of the science/mathematics analogy." (Maddy 1997, p. 131).

Though Maddy presents a very subtle and sophisticated argument around the disanalogies between science and mathematics in Chapters 5 and 6 of the part 3 of Maddy (1997), where she contrasts the realists' and the mathematical naturalists' approaches to the problem of the independent axioms of set theory, it is not necessary for us to look into the details. What is important is making it clear that Maddy's strategy in attacking the analogy between science and mathematics is the method of pinning down the relevant disanalogies. As is well-known, since this method is the standard one in evaluating analogical inferences, Maddy's confession has persuasiveness to a certain extent. If it is true that while natural science follows the principle of economy in introducing entities, mathematics follows the principle of maximization, this disanalogy must be certainly a relevant disanalogy powerful enough to force to give up the analogy between science and mathematics.

Maddy is, of course, fully aware of that fact that there were times in history when mathematics and physical science were inseparably intertwined. So, by emphasizing that situation changed enormously through the 19th century, and by dealing with modern mathematicsshe, wants to frame the discussion in such a way that her points would be more persuasive. Relying on Kline's general history of mathematics, Maddy points out that mathematicians increasingly began to introduce concepts that have minimal physical meaning. And, she presents some relevant examples such as that negative numbers and complex numbers were troublesome until they got geometrical interpretations, non-Euclidean geometries and the considerations of n-dimensional space yield ultimately pure mathematics completely freed from all spatial intuition, Cantor expanded the notion of number to transcendental number, abstract mathematics and the axiomatic method contributed to the emancipation of mathematics from the requirements of physics. And, Maddy claims that as a consequence of such a development of mathematics, "contemporary pure mathematics is pursued on the assumption that mathematicians should be free to investigate any and all objects, structures, and theories that capture their mathematical interest" (Maddy 1997, p. 210). In connection with the methodology of set theory, which is her main interest, she makes a very strong claim:

> If mathematics is to be allowed to expand freely in this way, and if set theory is to play the hoped-for foundational role, then set theory should not impose any limitations of its own: the set theoretic arena in which mathematics is to be modeled should be as generous as possible; the set theoretic axioms from which mathematical theorems are to be proved should be as powerful and fruitful as possible. (Maddy1997, pp. 210–211)

So, Maddy concludes that "the goal of founding mathematics without encumbering it generates the methodological admonition to MAXIMIZE". [ibid., p. 211]

3.2.5 Can the Analogy Between Science and Mathematics Be Abandoned?

Maddy's strategy to find from the maximizing principle the disanalogy between science and mathematics, thereby to reject the analogy between science and mathematics, and to announce ultimately the demise of mathematical realism seems clear enough by now. Thus, the central question of this paper, i.e., "why did Maddy give up mathematical realism, and arrive at irrealism or mathematical naturalism that have intransparent ontological stance?" has been solved. However, it is an another matter whether she was right in abandoning the analogy between science and mathematics. Indeed, some serious objections against Maddy in this regard have been filed. The most notables ones I have in mind are the criticisms raised by Baker and Dieterle (Baker 2004; Dieterle 1999). Insofar as their criticisms have to the point, we may even doubt whether Maddy's attempt to give up the analogy between science and mathematics itself is impossible.

To date, Maddy has not replied to Baker's incisive and contentful criticisms. Let us briefly introduce some of his criticisms. Contrary to Maddy's thought, Baker wants to show that the principle of maximization cannot draw the boundary line between science and mathematics. For this purpose, he takes the strategy to divide the issue into parts: the problem of the principle of maximization within set theory, the problem of the principle of maximization in mathematics other than set theory, and the problem of the principle of maximization in empirical science. Let me take some arguments from each of these. First, in set theory, Baker points out that the two aims, i.e., the aim to be "as generous as possible" and another aim to be "as powerful and fruitful as possible", do "not always pull in the same direction". [ibid.,] He also points out that as "mathematical axioms can expand ontology by making existence claims", "they can also contract ontology by imposing more stringent conditions for admission into the domain of a given theory". Furthermore, he perceptively indicates that "Restrictive axioms may turn out to be very fruitful, and powerful, despite (or maybe precisely *because*) they limit the set-theoretic arena" (Baker 2004, p. 271).

There is another interesting point noted by Baker. He believes that in the fields of mathematics other than set theory, especially in more applied areas, we find desires not to use unnecessarily strong mathematical theories for the given task. On the other hand, he is alert to note that there is no reason not to prefer expanded theory, in case there is no cost for using extra mathematical ontology. Consequently, Baker claims that in most of the areas of mathematics the role of the principle of maximization is "marginal at best" (Baker 2004, pp. 273–274).

The most intriguing point appears in Baker's criticism of Maddy's contrast of science and mathematics through the principles of parsimony and maximization. It is Baker's sharp observation that Maddy's contrast would be persuasive only when the principle of maximization does not work in empirical science. But in the case of theoretical physics, as testified by 20th century particle physics and quantum mechanics, the role of parsimony is not clear, unlike the situation in experimental physics, biology, and geology. Furthermore, according to Baker, the principle of maximization looms large as the essential part of the theoretical framework in these areas (Baker 2004, p. 274–276).

Though extremely interesting and important, it is obviously beyond the scope of this paper to delve in the controversy between Maddy and Baker further. Let it suffice to note that since it is controversial whether the principle of maximization, which was emphasized by Maddy as the crucial difference between science and mathematics, is truly peculiar to mathematics, Baker seems to contributing enormously by persuading us that it could have been too early for Maddy to abandon the analogy between science and mathematics.

On the other hand, Dieterle's criticism intends to show that Maddy's mathematical naturalism, which presupposes the disanalogy between science and mathematics, is bound to be led to some absurd consequences. Interestingly, Maddy (1997) anticipated Dieterle's criticism as a possible objection, and claimed that she can meet it ably. Dieterle now points out the insufficiency of Maddy's response to the criticism. The problem is this: Why should we give special treatment to mathematics in such a way that, unlike other non-scientific methods and practicies, mathematics is immune to any extrinsic criticisms? For example,

> Why, for example, can an astrologer not embrace astrological naturalism and defend that view with the same sort of arguments that Maddy uses to defend mathematical naturalism? (Dieterle 1999, 130; Maddy 1997, p. 203)

To this possible objection, Maddy responds that, unlike pluralists or relativists, scientific naturalists have good reasons to treat mathematics differently from other disciplines. One reason provides scientific naturalists with the ground to restrain criticism against mathematics. Nevertheless, another reason provides them with a strong motivation to explain mathematics. According to the former reason, science includes the entire spatio-temporal reality, i.e., the entirety of causal order. While pure mathematics has nothing to tell about this domain, astrology posits a new causal power, and gives novel predictions about the spatio-temporal events. On the other hand, even if we reinterpret astrology as dealing with supernatural power that does not interact causally with ordinary physical phenomena, the latter reason shows us that there is an important disanalogy betwee mathematics and astrology. Maddy refers to a simple reason that "mathematics is staggeringly useful, seemingly indispensable, to the practice of natural science, while astrology is not" (Maddy 1997, p. 204).

First, Dieterle observes that the two reasons invoked by Maddy are both amounting to disanalogies between mathematics and astrology. But the two disanalogies seem incompatible:

> But if mathematics is so useful in natural science, then it must have something to say about the spatio-temporal realm. And if mathematics literally has nothing to say about that realm, how can it be 'staggeringly useful, seemingly indispensable' to the practice of natural science? (Dieterle 1999, 132).

So, she finds a clue from the fact that "it would seem that the second disanalogy between mathematics and astrology is undermined by the first disanalogy" (Dieterle 1999, 132).

Next, she posits a fictional character called "theological naturalist", who "takes a stand on thelogy analogous to Maddy's stand on mathematics", and raise the following question:

> Given Maddy's arguments for mathematical naturalism, would she have grounds to reject theological naturalism? [ibid.]

It is not entirely clear for what purpose she introduces a theological naturalist. Of course, an important hint is given: Dieterle believes that it is the salient difference between non-causal astrologer and theological naturalist that, unlike the case of the former, "one can imagine that theology could be extremely (even staggeringly) useful in natural science". [ibid.]

Even then, however, it is still unclear whether Dieterle is guarding against the rhetorical power of Maddy's argument that illegitimately appeals to our disbelief in astrology, or, in converse, whether she aims herself such a rhetorical effect by appealing to the emotion of devoutly religious people.

Be that as it may, the arguments presented by Dieterle using theological naturalist seem very subtle and powerful. For example, Dieterle launches a dilemma to mathematical naturalist:

> In short, mathematics either is just a useful tool or it is not. If it is, then this may undermine the theological naturalism analogy, but it also undermines mathematical naturalism. All we end up with is scientific naturalism. If it is more than just a tool, then Maddy cannot say that mathematics has *nothing* to say about the domain of natural science, and we have no grounds to reject theological naturalism. Hence, regardless of whether mathematics is just a tool, we (as scientific naturalists) have no argument *for* mathematical naturalism that can *withstand* theological naturalism. At least not yet. (Dieterle 1999, 133)

As was the case with the controversy between Baker and Maddy, we are still waiting for Maddy's direct response to Dieterle in the controversy between them. If so, it would be too much to expect a definite answer in this article. My modest aim is simply to learn some morals from these debates. What is salient in Baker's criticism is that, by casting a doubt to the disnalogy between science and mathematics Maddy invokes in order to reject the analogy between science and mathematics, it seems established that it is too early to abandone the analogy between science and mathematics. On the other hand, Dieterle's criticism has interesting aspects in the way of arguing that focuses on the absurd consequences resulted by Maddy's mathematical naturalism. For, if Dieterle's argument is successful, not only the falsehood of Maddy's position would be confirmed, but also the conclusion that such a position is impossible from the start might be confirmed. Also, it is suggestive that both Maddy and Baker use arguments that are related to the analogies or disanalogies between science and mathematics. Now, encouraged by Baker's and Dieterle's attempts, I tend to think that, if Maddy's position is impossible from the outset, and the argument (i.e., the idea to abandon the analogy between science and mathematics due to the disanalogy between them) for the position is wrong, there is room for doubt that the task of establishing the argument itself is possible, at all. Of course, it is beyond the scope of this paper, by specifying such a doubt, to formulate an impossibility proof that shows the impossibility of abandoning the analogy between science and mathematics.

4 Concluding Remarks: What Maddy's Belief Change Means

In this chapter, I tried to find out why Maddy gave up her earlier position, i.e., mathematical realism. As a consequence, I made it clear that, contrary to Colyvan's view, it was not the result of giving up indispensability argument but her abandonment of the analogy between science and mathematics itself at a deeper and more wide-ranging level that was the crucial factor. From my point of view, it would have been nicer, if she elaborated her ontological stance in Maddy (1990), which I count to be a mathematical Aristotelianism (See Park 1994). Maddy's subsequent research turns out to be the total abandonment of the analogy between science and mathematics, and thereby realism. Such a move resulted in heading at a mathematical naturalism. As she characterize it as arealism, its ontological position is rather dubious, and I think all this development as her ontological regress caused by her unfortunate metaphilosophical speculation.[4]

References

Baker, A. (2004). Maximizing principles in mathematical methodology. *Logique et Analyse, 45,* 269–281.

Balaguer, M. (1994). Against (Maddian) naturalized platonism. *Philosophia Mathematica, 3,* 97–108.

Balaguer, M. (1996). Toward a nominalization of quantum mechanics. *Mind, 105,* 209–226.

Benacerraf, P. (1965). What numbers could not be. *The Philosophical Review, 74*(1), 47–73.

Benacerraf, P. (1973). Mathematical truth. *The Journal of Philosophy, 70,* 661–679.

Burgess, J. P. (2004). Comments on penelope Maddy's 'mathematical existence'. Unpublished comment read at Chapel Hill Colloquium (Oct. 2004); available at his homepage.

Burgess, J. P., & Rosen, G. (1997). *A subject with no object.* Oxford: Clarendon Press.

Cassou-Noguès, P. (2005). Goedel and 'the objective existence' of mathematical objects. *History and Philosophy of Logic, 26,* 211–228.

Chihara, C. S. (1982). A Goedelian thesis regarding mathematical objects: Do they exist? And can we perceive them? *Philosophical Review, 91,* 211–227.

Chihara, C. S. (1990). *Constructibility and mathematical existence.* Oxford: Clarendon Press; New York: Oxford University Press.

Chihara, C. S. (1998). *The worlds of possibility.* Oxford: Oxford University Press.

Chihara, C. S. (2004). *A structural account of mathematics.* Oxford: Oxford University Press.

Colyvan, M. (1998). In defence of indispensability. *Philosophia Mathematica, 6,* 39–62.

Colyvan, M. (1999). Contrastive empiricism and indispensability. *Erkenntnis, 51*(2–3), 323–332.

Colyvan, M. (2001). *The indispensability of mathematics.* Oxford: Oxford University Press.

---

[4]It is simply beyond the scope of Maddy's more recent work including Maddy (2007) and Maddy (2011). Insofar as her position in philosophy of mathematics is concerned, my discussion here seems to be sutatined. She has become more sensitive to metaphilosophical issues, which are not particularly relevant to my purpose. More pertinent wouls be Maddy (1996b, 1998a, 2000a, b, 2001a, b, 2005a, b).

Decock, L. (2003). Quine's weak and strong indispensability argument. *Journal of the General Philosophy of Science, 33,* 231–250.

Dieterle, J. M. (1999). Mathematical, astrological, and theological naturalism. *Philosophia Mathematica, 7,* 129–135.

Gödel, K. (1944). Russell's mathematical logic. In P. Schilpp (Ed.) *The Philosophy of Bertrand Russell* (Library of Living Philosophers (pp. 123–153). New York: Tudor, 1951. Reprinted in Gödel 1990, pp. 119–141.

Gödel, K. (1947). What is Cantor's continuum problem? *Amererican Mathematical Monthly, 54,* 515–525. (Reprinted in Gödel 1990, pp. 176–187).

Gödel, K. (1990). In S. Feferman, J. Dawson, S. Kleene, G. Moore, R. Solovay, J. van Heijenoort (Eds.), *Collected works. II: Publications 1938–1974.* Oxford: Oxford University Press.

Lee, J. (2005), *A study of mathematical realism based on the indispensability of mathematical entities in science* (Ph.D. Dissertation). Seoul: Dongguk University.

Leng, M. (2002). What's wrong with indispensability? *Synthese, 131,* 395–417.

Levine, A. (2005). Conjoining mathematical empiricism with mathematical realism: Maddy's account of set perception revisited. *Synthese, 145,* 425–448.

MacCallum, David. (2000). Conclusive reasons that we perceive sets. *International Studies in the Philosophy of Science, 14,* 25–42.

Maddy, P. (1980). Perception and mathematical intuition. *Philosophical Review, 89,* 163–196.

Maddy, P. (1984a). Mathematical epistemology: What is the question? *The Monist, 67,* 46–55.

Maddy, P. (1984b). How the causal theorist follows a rule. In H. Wettstein et al. (Eds.), *Causation and causal theories, midwest studies in philosophy* (Vol. 9, pp. 457–477). Minnesota: University of Minnesota Press.

Maddy, P. (1990). *Realism in mathematics.* Oxford: Oxford University Press.

Maddy, P. (1992). Indispensability and practice. *Journal of Philosophy, 89,* 275–289.

Maddy, P. (1994). Taking naturalism seriously. In D. Prawitz, B. Skyrms, D. Westerstahl (Eds.), *Logic, methodology and philosophy of science IX* (pp. 383–407). New York: Elsevier Science.

Maddy, P. (1995). Naturalism and ontology. *Philosophia Mathematica, 3,* 248–270.

Maddy, P. (1996a). The legacy of "mathematical truth. In A. Morton & S. Stich (Eds.), *Benacerraf and his Critics* (pp. 60–72). Oxford: Blackwell.

Maddy, P. (1996b). Ontological commitment: Between Quine and Duhem. In J. Tomberlin (Ed.), *Philosophical perspectives 10, metaphysics* (pp. 317–341). Cambridge: Blackwell.

Maddy, P. (1996c). Set theoretic naturalism. *Journal of Symbolic Logic, 61,* 490–514.

Maddy, P. (1997). *Naturalism in mathematics.* Oxford: Oxford University Press.

Maddy, P. (1998a). Naturalizing mathematical methodology. In M. Schirn (Ed.), *Philosophy of mathematics today* (pp. 175–193). Oxford: Oxford University Press.

Maddy, P. (1998b). How to be a naturalist about mathematics. In G. Dales & G. Oliveri (Eds.), *Truth in mathematics* (pp. 161–180). Oxford: Oxford University Press.

Maddy, P. (2000a). Naturalism and the a priori. In P. Boghossian & C. Peacocke (Eds.), *New essays on the a priori* (pp. 92–116). Oxford: Oxford University Press.

Maddy, P. (2000b). Does mathematics need new axioms? (symposium with Solomon Feferman, Harvey Friedman and John Steel). *Bulletin of Symbolic Logic, 6,* 413–422.

Maddy, P. (2001a). Naturalism: friends and foes. In J. Tomberlin (Ed.), *Philosophical perspectives 15, Metaphysics 2001* (pp. 37–67). Madlen, MA: Blackwell.

Maddy, P. (2001b). Some naturalistic remarks on set theoretic method. *Topoi, 20,* 17–27.

Maddy, P. (2005a). Three forms of naturalism. In S. Shapiro (Ed.), *Oxford handbook of philosophy of logic and mathematics* (pp. 437–459). Oxford: Oxford University Press.

Maddy, P. (2005b). Mathematical existence. *Bulletin of Symbolic Logic, 11*(3), 351–376.

Maddy, P. (2007). *Second philosophy: A naturalistic method.* Oxford: Oxford University Press.

Maddy, P. (2011). *Defending the axioms: On the philosophical foundations of set theory.* Oxford: Oxford University Press.

Moschovakis, Y. N. (1980). *Descriptive set theory.* Amsterdam: North Holland.

Park, W. (1994). Schools of contemporary philosophy of mathematics. *Science and Philosophy, 5,* 327–340 (Reprinted in Park, W. (1997), *In Search of Science Lost* (pp. 75–87). Seoul: Damron Sa. (in Korean).

Park, W. (2006). The ontological regress of Maddy's mathematical naturalism. *Korean Journal of Logic, 9,* 2. (in Korean).

Peressini, A. (1997). Troubles with indispensability: Applying Pure mathematics in physical theory. *Philosophia mathematica, 5,* 210–227.

Peressini, A. (2003). Review of Colyvan, the indispensability of mathematics. *Philosophia Mathematica, 11,* 208–223.

Pincock, C. (2004). A revealing flaw in Colyvan's indispensability argument. *Philosophy of Science, 71,* 61–79.

Putnam, H. (1971). *Philosophy of logic* (reprinted in Putnam (1979), 323-357).

Putnam, H. (1975). *What is a mathematical truth? (reprinted in Putnam (1979), Mathematics, Matter and Method (Philosophical papers, I),* (pp. 60–78). Cambridge: Cambridge University Press.

Quine, W. V. (1948). On what there is (reprinted in Quine, from a logical point of view, 2nd edn., Cambridge, Mass: Harvard University Press).

Quine, W. V. (1975). Five milestones of Empicisim (in Quine (1981), 67–72).

Quine, W. V. (1981). *Theories and things.* Cambridge, Mass: Harvard University Press.

Resnik, Michael D. (1995). Scientific versus mathematical realism: The indispensability argument. *Philosophia Mathematica, 3,* 166–174.

Rhee, J. (2003). *On the application of mathematics and its ontological implication* (MA thesis). Seoul: Ehwa Womans University.

Riskin, A. (1994). The most open question in the history of mathematics: A discussion of Maddy. *Philosophia Mathematica, 2,* 109–121.

Russell, B. (1906). On 'insoubilia' and their solution by symbolic logic (reprinted in Russell (1973, pp. 190–214)).

Russell, B. (1907). The regressive method of discovering the premises of mathematics (reprinted in Russell (1973, pp. 272–283)).

Russell, B. (1919). *Introduction to mathematical philosophy.* London: Allen and Unwin.

Russell, B. (1973). In D. Lacky (Ed.), *Essays in analysis.* London: Allen and Unwin.

Sober, E. (1993). Mathematics and Indispensability. *Philosophical Review, 102,* 35–57.

Tieszen, R. (1994). Review of penelope Maddy's realism in mathematics. *Philosophia Mathematica, 2,* 69–81.

Chapter 11
What if Haecceity Is not a Property?

Abstract In some sense, both ontological and epistemological problems related to individuation have been the focal issues in the philosophy of mathematics ever since Frege. However, such an interest becomes manifest in the rise of structuralism as one of the most promising positions in recent philosophy of mathematics. The most recent controversy between Keränen and Shapiro seems to be the culmination of this phenomenon. (See MacBride 2006b) Rather than taking sides, in this chapter, I propose to critically examine some common assumptions shared by both parties. In particular, I shall focus on their assumptions on (1) haecceity as an individual essence, (2) haecceity as a property, (3) the classification of properties, and thereby (4) the search for the principle of individuation in terms of properties. I shall argue that all these assumptions are mistaken and ungrounded from Scotus' point of view. Further, I will fathom what consequences would follow, if we reject each of these assumptions.

Keywords *Ante rem* structuralism · Haecceity · Identity of indiscernibles Individual essence · Jukka keränen · Stewart shapiro

1 Introduction

In some sense, both ontological and epistemological problems related to individuation have been the focal issues in the philosophy of mathematics ever since Frege. However, such an interest becomes manifest in the rise of structuralism as one of the most promising positions in recent philosophy of mathematics. The most recent controversy between Keränen and Shapiro seems to be the culmination of this phenomenon. There is no doubt that Keränen's challenge to *ante rem* structuralism contributed a lot to recent philosophy of mathematics. The fact that Shapiro withdrew some of his claims from his 1997 work, *Philosophy of Mathematics: Structure and Ontology* for unfortunately suggesting unsupportable implications is a sufficient ground for such a positive assessment (Shapiro 2006a, p. 133).

This chapter was published originally as Park (2016).

However, after their exchange of ideas in 2006 (Keränen 2006; Shapiro 2006a, b), there seems to be no significant progress in their controversy. Of course, unlike Keränen, who has been silent about their controversy since 2006, Shapiro expands the scope of the issues involved in several directions (Shapiro 2008a, b, 2012) shedding light on extremely interesting points related to mathematical practice. Insofar as the metaphysical aspects are concerned, however, they seem to be talking at cross purposes.

Rather than taking sides, in this chapter, I propose to critically examine some common assumptions shared by both parties. In particular, I shall focus on their assumptions on (1) haecceity as an individual essence, (2) haecceity as a property, (3) the classification of properties, and thereby (4) the search for the principle of individuation in terms of properties. I shall argue that all these assumptions are mistaken and ungrounded from Scotus' point of view. No defense, let alone proof, of Scotus' position will be attempted. For, I believe, it is meaningful in itself to cast doubt on some widely accepted assumptions. Further, I will fathom what consequences would follow, if we reject each of these assumptions. By casting doubt on the assumptions shared by Keränen and Shapiro, we may advance to a new stage of the controversy ignited by them. Insofar as this controversy centers around whether *ante rem* structuralism implies the identity of indiscernibles, the rejection of their common assumption of haecceity as individual essence must change its fortune drastically. The rejection of the assumption that haecceity is a property would have more destructive power. For, if we do not believe that haecceity is a property, it seems pointless to classify properties into general properties and haeccieties. As a consequence, we would have to rethink what is involved in our individuation task itself. In sum, I want to check whether we can expect more fruitful exchange of ideas about the identity of indiscernibles, thereby achieving progress in metaphysics in general as well as in the philosophy of mathematics, by criticizing the assumptions at issue.

I will proceed as follows. First, I shall show that both Keränen and Shapiro share all four assumptions in Sect. 2. Then, in Sect. 3, I will examine in what roles and functions these assumptions play in their controversy. John Duns Scotus' original view of *haecceitas* will be briefly presented in Sect. 4, only to cast a doubt against the assumptions. On that basis, I will fathom in Sects. 5 and 6 what consequences would follow from rejecting them.[1]

---

[1]Even though our discussion starts with the problems (possibly) peculiar to *ante rem* structuralism in philosophy of mathematics, it must be of interest to philosophers in other areas such as philosophy of quantum mechanics. In fact, it becomes more and more urgent to share results in all areas of philosophy, where individuation matters. Also, as identity of indiscernibles or *haecceitas* demonstrates, history of ontology could be instrumental for deeper understanding of our problems. In the same vein, the *ante rem* aspect of Shapiro's *ante rem* structuralism can be discussed in a broader historical context.

2   The Common Assumptions About Haecceity

Keränen launches an apparently fatal blow against Shapiro's *ante rem* structuralism by arguing that it "will either fail to present a tenable account of mathematical ontology or else must betray the very motivations from which it stems" (Keränen 2006, p. 146). According to him, any philosophical theory must provide us an account of identity by completing the blank in what he calls the 'identity schema': (IS) (Ibid., p. 147). He also claims that "all extant theories of ontology maintain that the identity of objects is governed by their *properties*" (Keränen 2001, 313). By assuming that these properties must be either a general property or a haecceity, he argues that the *ante rem* structuralist can adopt neither a general-property account nor a haecceity account. If the *ante rem* structuralist grasps the first horn, that will result in the identity problem. On the other hand, if he grasps the second horn, that will conflict with the spirit and motivations of *ante rem* structuralism (Ibid. 314). Clearly, Keränen subscribes to all four assumptions: (1) that haecceitas is an individual essence, (2) that haecceity is a property, (3) that properties are classified into general and haecceitic, and thereby (4) that the search for the principle of individuation should be done in terms of properties.

Let us take a closer look at Keränen's acceptance of these four assumptions. The case of assumption (1) is straightforward, for he defines haecceity as "a property that can be possessed by one entity alone" (Ibid. 313). As an example, he presents '*being identical to a*', and claims that "'haecceity of a' is often cashed out in this way" with referring to Adams (1979). Given assumption (1), it is all too natural to accept the assumption (2). For, an individual essence, which is a property that can be possessed by one entity alone, must be a property. If assumption (2) is accepted, the case of assumption (3) is also obvious, for Keränen made his claims about assumption (2) in the same context where he classified properties into general and haecceitic:

> To the best of our knowledge, all extant theories of ontology maintain that the identity of objects is governed by their *properties*. To put it in the most general terms possible, '*a = b*' is true if and only if for all properties φ in some class Φ, $\varphi(a)$ if and only if $\varphi(b)$. Depending on what sort of properties the theory T takes the class Φ to include, the account of identity it gives will be either a *general-property account* or a *haecceity account*. A general property is a property that can be possessed by more than one entity, whereas a haecceity is a property that can be possessed by one entity alone. (Keränen 2001, 313)

The case of assumption (4) may be a bit less straightforward compared to the assumptions (1), (2), and (3), for Keränen here merely reports an alleged fact that "all extant theories of ontology maintain that the identity of objects is governed by their *properties*" (Ibid.; see also Keränen 2006, p. 154). He clearly advances his ideas, however, by simply adopting this alleged fact. As a result, we may safely view him as accepting assumption (4).

Also, there is no doubt that Shapiro at least implicitly adopts all these assumptions. Let us find some examples, which clearly show that Shapiro employs assumption (1). Shapiro draws our attention to the fact that some philosophers resort to haecceities to discharge individuation tasks:

Some philosophers have posited that each object has an 'haecceity', which is a property enjoyed by it and it alone. For those philosophers, the individuation task is trivially discharged, by filling in the blank in (IND) with a term for the haecceity for a:

$$\forall x(x = a \equiv Ha(x)),$$

Or perhaps just

$$\forall x(x = a \equiv x = a). \text{ (Shapiro 2006a, b, p. 133)}$$

Here Shapiro clearly counts haecceity as 'a property enjoyed by an object and it alone'. That means, he understands it as an individual essence. The case of assumption (2) is also evident. For, in accepting assumption (1), he already views it as a property. Similarly, when Shapiro claims that "presumably, a structuralist cannot accept haecceities for places, since a haecceity seems to be a non-structural property", he already treats it as a property (Ibid.).

In view of the passages quoted in the previous paragraph, it is also evident that Shapiro implicitly accepts assumption (3), as Keränen does. The only difference is that, instead of the dichotomy of general and haecceitic properties, Shapiro uses that of structural and haecceitic properties. Probably, the following quote from Shapiro (2008b) may show this point a bit more clearly:

> Why think that our languages are rich enough? If each mathematical object has an haecceity, a property that applies to it alone, then the job of individuation is done trivially, but at least it is done. The existence of the haecceity of an object a provides the fact that makes a the object it is, distinct from any other. Only a has *that* particular haecceity. The problem, of course, is that since it is virtually analytic that haecceities are not structural properties, the *ante rem* structuralist cannot invoke this trivial resolution of the individuation task. (Shapiro 2008b, 288)

As is the case with Keränen, assumption (4) seems never explicitly mentioned by Shapiro. In view of the fact that Shapiro never considers something other than structural or non-structural properties as candidates for individuation tasks, however, it may be quite safe to conclude that he also adopts assumption (4) implicitly. At least, Shapiro has no qualms with advancing the discussion about individuation in terms of properties: "If the individuation is to be done with properties or propositional functions, regarded as objective and independent of language, then why think that the intensional realm of properties and propositional functions matches up to the realm of objects in the relevant way?" (Shapiro 2006a, p. 137).

3   Haecceity and the Identity of Indiscernibles

Now it is time to take a closer look at how these assumptions shared by Keränen and Shapiro play significant roles and functions in their disputes. For this purpose, the status of the identity of indiscernibles looms large. This is evident even from the introductory remark of Keränen (2001):

> In this paper, we will argue that in spite of its many attractions, realist structuralism must ultimately be rejected. On the one hand, we will show that mathematical structures typically

contain places that are *indiscernible* if individuated solely by the relations they have to one another. On the other hand, we will argue that any account of place-identity available to the realist structuralist entails that relationally indiscernible places are *identical*. Since she maintains that mathematical singular terms denote places in structures, the realist structuralist is therefore committed to saying that, for example, $1 = -1$ in the group of integers $(Z, +)$. We will call this predicament the *identity problem*. We will argue that the identity problem vitiates the realist structuralist account of ontology beyond repair and conclude that *nominalism* is the only potentially viable version of structuralism. (Keränen 2001, 308)

Apparently, Keränen is here launching a dilemma for an *ante rem* structuralist, and its force seems to be widely appreciated.[2]

In his criticism of an *ante rem* structuralist, it is crucial for Keränen to argue that she is committed to the identity of indiscernibles. Keränen claims that a structuralist should complete (IS) as in the following formula:

$$(STR)\ \forall x\,\forall y\,(x = y \Leftrightarrow \forall\varphi\,(\varphi \in \Phi \Rightarrow (\varphi(x) \Leftrightarrow \varphi(y)))).$$

(Keränen 2001, 316; Keränen 2006, p. 147)

Further, he claims that (STR) "entails a (rather strong) form of the 'Leibniz principle' of the identity of indiscernibles" (Keränen 2001, 317, n. 16; Keränen 2006, p. 148).

Shapiro is fully understanding this, as is clear from his apt presentation of the problem for *ante rem* structuralist:

> The problem, he [Keränen] says, is that the ante rem structuralist is committed to the identity of indiscernibles, and this forces the structuralist to *identify* all of the places of each finite cardinal structure with each other, so that each structure has only one place after all. The ante rem structuralist must also identify the complex numbers i and $-i$, which contradicts the theorem that each number (other than 0) has two *distinct* square roots. (Shapiro 2006a, p. 132)

It is also crucial for Shapiro to deny that *ante rem* structuralist is committed to the identity of indiscernibles. As it turns out, however, Shapiro has apparent difficulties for such a denial. Let us try to understand what difficulties are involved.

Shapiro concedes that "Keränen correctly derives the identity of indiscernibles from the above thesis (STR)", so he apparently faces a serious difficulty. Since Shapiro further agrees that "(STR) entails that there is a non-trivial resolution of the individuation task", he has to concede that "there can be no non-trivial resolution of the individuation task" for *ante rem* structuralists. As Shapiro concludes, *ante rem* structuralists "must reject (STR)", but how?

---

[2]See, for example, McBride (2006a, 64), MacBride (2005, 219), and Ladyman (2005, 219) for different formulations of the alleged dilemma. It is not clear which way of formulating the dilemma is the best, but it seems that Keränen and Shapiro are more careful than others at least in that they are more sensitive to the possible roles and functions of identity of indiscernibles in this alleged dilemma. Here I am indebted to an anonymous reviewer for more focused discussion.

According to Shapiro (2006b), he "proposed a modus tollens at this point, rejecting (STR), and, in particular, rejecting the provisos on the set Φ" (Shapiro 2006b, p. 166). It is indeed correct that Shapiro's basic strategy to respond to Keränen's criticism was to attack identity of indiscernibles in order to reject (STR), but it is an entirely different matter whether he was successful in executing that plan in Shapiro (2006a). There is no doubt that Shapiro expressed his antipathy against identity of indiscernibles. For example, he wrote:

> For what it is worth, I have never been moved by the identity of indiscernibles. There are two ways that this thesis about identity can be understood, and both are contentious (at best). (Shapiro 2006a, p. 136)

Also, he wrote:

> What reason is there to think that the realm of properties and propositional functions is up to the task of individuating each and every object? Unless, of course, there are haecceities, in which case the identity of indiscernibles is trivially true, and not very interesting. If there are not enough properties to individuate each object, then there will be distinct objects that share all their properties. (Shapiro 2006a, p. 137)

Much more salient in Shapiro (2006a) is how anxious he was to deny "ante rem structuralist is committed to a crucial premise for the identity of in discernible, and for Keränen's principle (STR)" (Shapiro 2006a, p. 140). In a similar vein, one might feel that Shapiro was somewhat too defensive, when he asked whether the individuation task is "a reasonable demand on a philosophical or scientific theory" (Shapiro 2006a, p. 134). In fact, this presents Keränen a perfect opportunity to clarify what are required for the individuation task. In view of Keränen's clear distinction between (1) an account of identity, (2) a condition of identity, and (3) explicit display, and an apt example using Zermelo-Frankel set theory, the individuation task appears to be quite a reasonable demand, contrary to Shapiro's complaint (Keränen 2006, pp. 150–1).

Be that as it may, as we saw above, in both Keränen's criticism of *ante rem* structuralism and Shapiro's defense of it, haecceity already plays very important roles and functions. However, it is Keränen's reply to Shapiro (Keränen 2006), where we can witness the most extensive discussion of haecceity. In fact, I wonder whether there is any other single philosophical text that invokes haecceity so frequently. Why is haecceity so important for Keränen in his dispute with Shapiro? Why does he have to discuss haecceity so extensively? Why has haecceity become more and more salient as the dispute proceeds?

Contrary to appearance, it is not difficult to find answers to these questions from Keränen's own words. Keränen set the trap for *ante rem* structuralist by exploiting certain peculiarities of haecceity: *ante rem*structuralist must be uncomfortable, if not unable, to appeal to haecceity for individuation task. To a certain extent, Keränen was right in his anticipation for Shapiro's response. The very structure of Shapiro's response turns out to be around the distinction between "what he [Shapiro] calls a 'trivial' and 'non-trivial' way of completing the blank in the identity schema" (Keränen 2006, p. 149). Needless to say, the so-called trivial way is nothing but appealing to haecceity, and, that is of utmost importance for this analysis. In passing, we may also note here that inadvertently Shapiro seems selecting the moves exactly as was anticipated by Keränen. Insofar as Shapiro also accepts the assumptions mentioned above, probably he does not have any other option. Be that as it may, Keränen identifies "four principal components" in Shapiro's response to his criticism. The first is what Keränen call the 'trivializing objection', i.e., the idea that it is "too much to ask of *any* theory of (mathematical) ontology" to provide us with a non-trivial way of completing the identity schema. The second is to deny that *eliminative* structuralism is immune to the identity problem. The third is the attack on the alleged Leibniz Principle. The final component is Shapiro's attempts to withdraw some of his earlier claims (Ibid.). In one way or another, I will touch upon each of these four components. But, for my present purpose, we need to focus on the identity of indiscernibles.

Probably, the most challenging aspect of Keränen's position is that he claims the identity of indiscernibles to be true. Against Shapiro, Keränen claims that "it is beside the point whether the Leibniz principle is interesting; the only relevant issue is whether it is *true*". Then, he presents his own attitude about the Leibniz principle as follows:

> I believe that given any domain of objects, there is *some* fact that metaphysically under-writes the distinctness of any two distinct objects in that domain.* I will argue ... that that feature must be analyzable in terms of properties of individual objects. Hence, I believe that given any domain of objects, some formula of the form
>
> $$\forall x \, \forall y \, (x = y \Leftrightarrow \forall \varphi \, (\varphi \in \Phi \Rightarrow (\varphi(x) \Leftrightarrow \varphi(y)))),$$
>
> is true of that domain—a form of the Leibniz principle. (*Keränen's own footnote: see Keränen 2001, 312–13) (Keränen 2006, 154–5)

As was hinted at by Shapiro, the identity of indiscernibles has been counted to be controversial, to say the least. Its converse, i.e., the indiscernibility of identicals has been accepted universally as a principle of logic and called "Leibniz's law". Keränen intentionally calls the conjunction of the identity of indiscernibles and indiscernibility of identicals "Leibniz principle", championing the truth of identity of indiscernibles. We may note that Keränen is extremely ambitious and contentious in such a move.

That, however, is not the end of the story. Rather, the most radical and challenging thought of Keränen on the identity of indiscernibles is found in his attempt to demonstrate that "Shapiro has no choice but to adopt Haecceitism" (Ibid.). Let me quote in full a long paragraph from Keränen (2006), which amounts to a most intriguing *reductio ad absurdum* of haecceitism:

> Suppose that you think that the objects *a* and *b* are essentially indiscernible and yet distinct; you must still think that there is *something* about the world that is responsible for the objects being two and not one. Suppose that you think that the non-essential properties of *a* and *b* are not up to the task; you must still think that there is *something* about the world that is. Suppose that you think that it is a primitive, 'brute' fact that *a* and *b* are two objects there rather than one; surely you must still think that there is something about *each one* that makes it the case that it is the object it is, and not the other. For, surely you want to be able to make sense of each object *being identical to itself*. Note that we cannot say that it is the very *same* property of 'being identical to itself' for then we would again have no

metaphysical foundation for saying that the two objects really are two and not one. Thus, we must understand the description of this property 'indexically', so that each object has a *unique* property *being identical to itself*. But, surely the property *being identical to itself* is an essential property of an object if anything is. It follows trivially that

(b) $\forall x \, \forall y \, (x = y \leftarrow \forall \varphi \, (\varphi E(x) \Leftrightarrow \varphi E(y)))$.

Thus, if you reject (b), you are committed to adopting (b), a contradiction. (Keränen 2006, pp. 156–7)

By this intriguingly subtle argument, Keränen intends to demonstrate that "there are no distinct but essentially indiscernible objects" (Keränen 2006, p. 156). As it stands, the first supposition that the objects *a* and *b* are essentially indiscernible and yet distinct, which amounts to rejecting (b), must be the assumption for the *reductio ad absurdum*. From that supposition we derived (b), so, the announcement of having secured a contradiction at the end must be the fanfare celebrating the success of the attempted *reductio ad absurdum*.

Though this argument is quite subtle and interesting, there must be certain weaknesses in it. One might question whether the idea that "each object has a unique property being identical to itself" has been correctly formalized in (b). Further, one might find fault with the preoccupation with "the property of being identical to itself". What is impressive in this argument is rather the fact that here and there Keränen shows a very sincere and serious attitude to understand the opponents' position. All the subsidiary suppositions introduced consecutively in the first half of the argument seem to do justice to what the opponents actually believe and subscribe to. In fact, such opponents' positions might approach quite closely to Scotus' original concept of *haecceitas*. I will return to this point later.

4   *Haecceitas* from Scotus' Point of View[3]

As is well-known, the notion of haecceity was introduced by John Duns Scotus (c. 1265/6–1308), who is arguably one of the most important and influential philosophers in the High Middle Ages. As his nickname "the Subtle Doctor" indicates, he left unmistakable mark in all major fields of philosophy. Especially as a metaphysician, he is famous for his discussion of common nature theory of universals, and haecceity theory of individuation (Scotus 1992).[4] If so, I think, we have enough reason to doubt whether the fairly casual reference to haecceity found in Keränen and Shapiro is justifiable. This could be a serious problem, since Scotus'

---

[3]This section is drawn from Park (1988, 1990, 1998).

[4]Cross (1999) provides us with a nice introduction to Scotus' philosophy. Probably, the most recent and extensive treatment of Scotus' notion of haecceity is Andrews (2010). See also Cross (2003, 2014).

notion of haecceity has been widely conflated with individual essence (Park 1990).[5] In some sense, Keränen and Shapiro have quite understandable reasons to follow the tradition that equates haecceity with individual essence. It seems to be Leibniz more than anyone else who is to blame for this conflation. Indeed, Leibniz also had some very good reasons for all this. It is not a mere historical accident that Leibniz seems most responsible for all subsequent controversies related to the principle of the identity of indiscernibles and haecceity (Park 1998).

In his treatise on the problem of individuation, Scotus rejected a nominalistic theory according to which a material substance is of its nature singular and individual (Scotus 1992). If a theory of individuation by individual essences is such a theory, as I believe it to be, it would be absurd to interpret *haecceitas* as an individual essence. Since *haecceitas* is not definable and it is unlikely that we are acquainted with it, its defense must be based on very substantial arguments. Indeed, such arguments are found in Scotus. He proposed two arguments which jointly demonstrate that there must be something positive in the category of substance that individuates the specific nature. Especially, Scotus' criticism of the nominalistic position amounts to a major part of the first argument in his two arguments for postulating *haecceitas*. In other words, if an individual essence were the principle of individuation, and as a consequence the material substance were of its nature this or that individual, then there would be no need for postulating *haecceitas*. Thus, *haecceitas* cannot be an individual essence for Scotus.

His first argument [henceforth argument #1] is based on the idea that unity always follows upon certain entity; it aims to show that something positive is necessary for the individuation of the specific nature. His second argument [henceforth argument #2] is based on the idea that the difference between two things is ultimately reduced to something primarily diverse whereby they differ; it aims to show that the positive something in question should belong to the category of substance. Let me quote these two arguments first:

> [Argument #1] My first point then is to show that the specific nature is contracted to this singularity by something positive. All unity follows upon some entity, –and just as 'one' in general follows upon 'being' in general, so analogously with what is special. Hence the unity characteristic of singularity, which excludes any sort of division, will have some analogous entity as its base. But such unity does not stem from the entity of the nature. For, as was shown above [nn. 9-25], the unity of nature is less than *this* singularity that is numerical unity. That is why the unity characteristic of nature can stand in opposition to this [numerical] unity and is not a sufficient reason for such. Hence this must stem formally

[5]Allan B. Wolter cites the following items as examples of such a conflation: Bergmann (1964, pp. 160, 165, 287; Bergmann 1967, pp. 167, 191, 199, 204, 222; Plantinga 1978, 132; Chisholm 1986, 160; Losonsky 1987, 253) (Wolter 1992, xix, n. 26; Park 1990, 377, n. 8). Further example would be Rosenkrantz (1993). One anonymous reviewer counts my observation that 'haecceity has been widely conflated with individual essence' as correct. Nevertheless, the reviewer thinks it irrelevant, for "Shapiro and Keränen are not concerned with individual essence so there is nothing for them to conflate with". My response should be obvious: insofar as they assume haecceity as a property, it is their burden to indicate what kind of property that property is. Please see note 11 below for further discussion of connected points.

from some extra entity besides what is essential to the specific entity. – This unity then does not follow from the specific entity. Nevertheless that entity [or haecceity] from which it does stem, forms a *per se* unity with the specific nature, because the individual – as was proved above [nn. 65, 72-76, 87, 91-93] – is a *per se* unity and not through unity of another genus [such as that of quantity].It follows then that the specific nature is determined to be this individual by something positive. (Scotus 1992, pp. 81–3; Dist. 3, q. 6, n. 166)

[Argument #2] Furthermore, things that differ are "other-same things"[6]; but Socrates and Plato differ; hence there must be something whereby they differ, the ultimate basis of their difference. But the nature in the one and the other is not primarily the cause of their difference, but their agreement. Though the nature in one is not the nature in the other, nature and nature are not that whereby the two differ primarily, but that whereby they agree (for they do not differ just of themselves – otherwise there would be no real agreement between them), hence there must be something else whereby they differ. But this is not quantity, nor existence, nor a negation, as was established in the preceding questions [nn. 153-163]; therefore, it must be something positive in the category of substance, contracting the specific nature. (Ibid., p. 83; n. 167)

For my present purpose, it is argument #2 that is crucially important. And argument #1 is needed only insofar as it establishes one of the premises of argument #2. In order to see this point, argument #2 may be reconstructed as follows:

1. When things which have the same nature differ, there must be something whereby they differ,
2. The nature in things that differ is not the cause of their difference but of their agreement.

3. [Therefore]
There must be something other than the nature whereby they differ. (by 1 and 2)
4. And that by which they differ is either quantity or existence or negation or something positive in the category of substance contracting the specific nature.
5. But that by which they differ is not quantity, nor existence, nor negation. (by Scotus' discussion in previous questions)

Therefore,
6. It must be something positive in the category of substance, contracting the specific nature. (by 4 and 5)
QED.

As is clear in the reconstruction, this argument is, in fact, a sorites composed of two sub-arguments. But the guiding idea of Scotus' reasoning is clear enough. Only lines 1 and 4 need some comments. For we need to know how they are established. Line 1 was established in argument #1. But the case of line 4 is different. Let us not

---

[6]In Latin, it reads "diversaaliguid-idem entia". Cf. Metaphysics V, ch. 9, 1018112-13.

forget the point of argument #2. It is to prove that positive something whereby individuals in a species differ is in the category of substance. So, the success of this proof depends on whether line 4 is really exhaustive in enumerating possibilities. If so, and line 5 is justified, then the conclusion should be accepted. Scotus must think that line 4 is trivially true because he suppressed it. On the other hand, line 5 requires justification, which Scotus gave in questions 2 through 5.

Given Scotus' arguments #1 and #2, which were meant to prove jointly that there must be something positive in the category of substance which individuates the specific nature, now we can understand better not only the arguments but also the overall structure of Scotus' treatise on individuation. Scotus' arguments for postulating *haecceitas* found in question 6 are merely the summary of what he had been doing in questions 2 through 5. Argument #1 is a summary of question 2, where he concludes that a material substance is individuated by something positive. And argument #2 is a summary of what he did in questions 3 through 5. So, even though there seem to be only two short arguments for postulating *haecceitas*, actually they are very strong, because they are in fact built of blocks which were carefully prepared one by one in extensive previous discussions.

For those who might think that Scotus' criticism of the nominalist view does not entail that there is no individual essence in the modern sense, we may invoke Scotus' triple analogy between an individual difference (*haecceitas*) and a specific difference (Dist. 3, q. 6, nn. 170-2).[7] If we consider the possible consequences of substituting "individual essence" for "individual difference" in Scotus' analogy, we get absurd results. For example, in the third way, we would have to say that any two individual essences are "primarily diverse differences". As long as they share at least one characteristic, however, individual essences cannot be ultimately different. Thus, *haecceitas* (individual difference), in the sense of Scotus, cannot be an individual essence.[8]

5   What if Haecceity Is not a Property?[9]

If *haecceitas* is not individual essence at least for Scotus, there is already a very good reason to cast a doubt for assumption (2) that haecceity is a property. Assumption (3), i.e., that all properties are classified into general and haecceitic,

---

[7]Scotus compared an individual and a species in terms of their relationship to what is below each, to what is above which, and to what is on a par with each (Dist. 3, q. 6, nn. 170-2; Park 1990).

[8]This sketch is drawn from Park (2000). A fuller version is presented in Park (1990). Park (1990), entitled as "*Haecceitas* and the Bare Particular" paid much more attention to refuting philosophers such as Jorge J. E. Gracia, who tend to assimilate *haecceitas* to bare particulars, than those who assimilate *haecceitas* to an individual essence.

[9]I am indebted to an anonymous reviewer, who suggested Gracia (1996) as anticipating my thesis that *haecceitas* cannot be an essence. See especially pp. 240–244, where Gracia enumerated Scotus' negative descriptions of *haecceitas*. McMichael (1983) is another rare example. According to him, Allan Bäck has told him "that Scotus' haecceities are probably not properties" (p. 59, n. 14).

could be also potentially troublesome. If Keränen and Shapiro mean by "haecceity" nothing but "individual essence", as I believe them to do, such a classification would become either non-exhaustive or pointless. It would be non-exhaustive, if haecceity (*haecceitas* in the sense of Scotus) is a property different from individual essence (haecceity in the sense of Keränen and Shapiro). It would be pointless, if haecceity is not a property. In order to make it explicit, we need one more premise, i.e., that individual essence is a general property. As a matter of fact, Scotus implicitly assumed exactly this point in the first way of his triple analogy. For, if we substitute "individual essence" for "individual difference (*haecceitas*)" in the analogy, we have to say that "the individual essence even excludes division into further individuals of the same sort". If haecceity is not a property, and if even individual essences are in fact general, as Scotus and I believe them to be, then the dichotomy of general and haecceitic properties would be pointless.

Some people might think that my reasoning in the last paragraph is unnecessarily convoluted. For, haecceity (in the sense of Keränen and Shapiro) is just a property from their point of view. As a matter of fact, it is not difficult to find examples. Let us consider Rosenkrantz's attempts to define haecceity and individual essence:

(D1) F is a haecceity = df. $(\exists x)$ (F is the property of being identical with x.)

(Rosenkrantz 1993, p. 3)

(D2) F is a haecceity = df. F is possibly such that: $(\exists x)$ (F is the property of being identical with x.) (Ibid., p. 21)

(D3) E is an individual essence = df. $(\exists x)$ (x necessarily exemplifies E, and E is necessarily such that $\sim(\exists y)$ ($y \neq x$ & y exemplifies E.)) (Ibid., 42)

(D4) E is an individual essence = df. E is possibly such that: $(\exists x)$ (x necessarily exemplifies E, and E is necessarily such that $\sim(\exists y)$ ($y \neq x$ & y exemplifies E.))

(Ibid., p. 43; See Ujvári 2013 for more recent discussion)

I simply fail to see any difference between haecceity in the sense of Rosenkrantz and individual essence. According to him, however, even if "every haecceity is an individual essence", "some individual essences are not haecceities". Be that as it may, what is important for my present purpose is that Rosenkrantz just treats haecceity as a property in his definitions. So, insofar as haecceity is counted to be a property by my opponents, the only thing I need to show must be that if haecceity (in the sense of Scotus) is not a property, the dichotomy of properties into general and haecceitic (i.e., individual essence) is pointless.

Now I propose to consider the possibility that haecceity is not a property,[10] however implausible or awkward it may sound. I think that this possibility would enable us to see some crucially important parts of both Keränen's and Shapiro's arguments from an entirely different angle, ultimately advancing to a new stage of the controversy ignited by them. One obvious consequence of challenging the assumptions shared by Keränen and Shapiro, as suggested above, is that some of

---

[10]Please note that by doing this we are thereby fathoming at the same time the possibility of understanding individuation not in terms of properties.

the crucially important claims made by them would not only lose their cutting edge but also have entirely different unintended meaning. As we saw above, for example, Shapiro wrote:

> Unless, of course, there are haecceities, in which case the identity of indiscernibles is trivially true, and not very interesting. (Shapiro 2006a, p. 137)

Then, Keränen replied that

> it is beside the point whether the Leibniz principle is interesting; the only relevant issue is whether it is *true*. (Keränen 2006)

Now, it should be clear that, only when we understand them as meaning "individual essences" by "haecceities" in Shapiro's claim, both Keränen's and Shapiro's claims make sense.[11] Then, let us understand "haecceities" in Shapiro's claim as meaning "*haecceitas* (in the sense of Scotus)" in order to see whether Keränen's and Shapiro's claims would still make sense. In such a case, identity of indiscernibles would not be trivially true. Nor would it be uninteresting. Obviously, this would be uncomfortable to both Keränen and Shapiro.[12]

If I am on the right track in challenging assumption (2) that haecceity is a property, there are many other potentially important consequences. If haecceity is not a property, it is impossible to have formalization such as "$\forall x\ (x = a \equiv Ha(x))$" or "$Hx = Hy$" (Shapiro 2006a, p. 133). In other words, we should not misunderstand expressions like "haecceity of Socrates" as referring to a unique property possessed by Socrates only.

Nor could we use the plural form such as "haecceities". For, in the sense of Scotus, "haecceity that individuates a certain individual substance" and "haecceity that individuates another one" are simply others. Keränen, Shapiro, and

---

[11]At this stage, one anonymous reviewer complained that "none of the author's quotations either from Shapiro's texts of Keränen's texts…show that Shapiro and Keränen use the expression 'individual essence'—either explicitly or by implicit reference!" (Emphasis is the reviewer's) Based on this observation, the reviewer presented a very subtle defense of Shapiro and Keränen: "The *denial* of individual essence does make sense within the context of Scotus since he is concerned with the contraction of the specific nature. But the *assertion* of individual essence (projected onto Shapiro and Keränen by author) does not make sense from the point of view of Shapiro and Keränen". (Emphasis is the reviewer's) Instinctively, I would like to respond by reminding them that it is their burden to indicate what kind of property haecceity is. If it is not an individual essence, what kind of property do they have in mind in assuming haecceity as a property?

[12]One anonymous reviewer criticizes incisively my failure to provide a link between the first and the second part of this paper. The only consolation is that the same reviewer finds from my discussion in this paragraph "a hint at the direction for getting a new impetus from revisiting the 'history of ontology'". The reviewer even raises a series of insightful questions to answer in that direction: "can distinct but property-indiscernible mathematical objects be accounted for in terms of some haecceitistic property? If yes, how could it be taken as non-trivial? Can Scotus's alternative reading of *haecceitas* be of any help here?" In the next section, I will try to give at least some partial responses to these questions.

Rosenkrantz may object to this, by pointing out that they also capture and do justice to the simple diversity of haecceities. For, the property *being identical to itself* cannot be shared by many. Remember that Keränen put this point nicely, when he wrote that "each object has a *unique* property *being identical to itself*". But, he also claimed immediately that "surely the property *being identical to itself* is an essential property of an object if anything is" (Keränen 2006, pp. 157). Evidently, they fall victim to a kind of equivocation: *being identical to itself* is sometimes understood as an un-sharable property, and at other time as a sharable property. It is rather tempting to launch a dilemma for them at this moment. If *being identical to itself* is un-sharable, it is not a property; but if it is sharable, it is just a general property. The only way out from the dilemma seems to be grasping the first horn, thereby abandoning their assumption that haecceity (in their sense) is a property.

What is remarkable to note here is the fact that Scotus seems to anticipate all these possible problems involved in the speculation about haecceity. For, at the beginning of his treatise on the principle of individuation, he makes it crystal clear that in raising the question "whether a material substance by its very nature is a 'this', that is, singular and individual", we should not understand 'singularity' as a second intention (Scotus 1992, p. 3). *Haecceitas* (haecceity in the sense of Scotus) is what individuates this this. It is not a property unique to this such as *being identical to itself*.

Finally, as a consequence of our discussion challenging the assumptions (1), (2), and (3), we may also cast a doubt on the assumption (4), i.e., Keränen's and Shapiro's understanding of individuation task exclusively in terms of properties.

6 Concluding Remarks

All assumptions shared by Keränen and Shapiro, i.e., (1) haecceity as an individual essence, (2) haecceity as a property, (3) the classification of properties, and thereby (4) the search for the principle of individuation in terms of properties, are shown to be ungrounded from Scotus' point of view. As a consequence, it seems urgent to examine what consequences would follow, if we reject each of these assumptions. Though I discussed a few of them in the previous section, it is by no means enough for appreciating the full implication of those consequences to the future direction of mathematical structuralism. What is needed here, I think, is to examine how those consequences might affect Shapiro's original motivation to champion *ante rem* mathematical structuralism.[13]

---

[13]Virtually all anonymous reviewers require such a discussion, even though it is simply beyond my ability to show the future directions of mathematical structuralism. The discussion below would have been impossible without one reviewer, who specifically suggested to touch upon theories of essences.

Shapiro believes that two traditional views stand out throughout the long history of the problem of the nature and status of universals:

> One, due to Plato, is that universals exist prior to and independent of any items that may instantiate them....This view is sometimes called "*ante rem* realism," and universals so construed are "*ante rem* universals." The main alternative, attributed to Aristotle, is that universals are ontologically dependent on their instances....Forms so construed are called "in re universals," and the view is sometimes called "in re realism." (Shapiro 1997, p. 84)

By lumping together other alternate views like conceptualism and nominalism with in re realism, Shapiro makes explicit his belief that "the important distinction is between *ante rem* realism and the others" (Ibid., pp. 84–85). Among the three options for mathematical structuralism, i.e., ontological eliminative structuralism, modal eliminative structuralism, and *ante rem* structuralism, he opts for the *ante rem* option. The main attraction of this position must be that it "adopts an *ante rem* realism toward structures", whether it is "the most perspicuous and least artificial of the three" (Ibid., pp. 87–90).

Now, the discussion of expressions such as "*ante rem*" and "in re" in the context of discussing the problem of universals clearly reminds us of Avicenna's celebrated claim for the *triplex status naturae*: i.e., "as existing in the things (in the individuals), or as existing in the mind, *or as independent of any sort of existence*" (Angelelli 1967, p. 143).[14] We seem to have a reason to be quite curious about why Shapiro does not invoke Avicenna in discussing *ante rem* option. We could be more curious, if we are informed of the fact that, though strongly neo-platonized, Avicenna was still working in the Aristotelian context, as were the neo-platonic Greek commentators of Aristotle (See Lloyd 1981, pp. 62–66). The point is that there might be a room for an Aristotelian to embrace *ante rem* aspects of universals. Interestingly, Scotus seems to be the actual example of an Aristotelian adopting such a position by his famous theory of *natura communis*.

Joseph Owens aptly reconstructs the situation faced by a thirteenth century medieval Latin philosopher against the Avicennian background as follows:

> He would either have to accept the denial of unity in the essence at its face value and neglect the Avicennian teaching on the proper being of that essence, or else he would be forced to allow the essence its own proper being or entity and so would have to set up a unity corresponding to that entity, qualifying or explaining away the Avicennian assertions that unity does not apply to the nature taken as such. (Owens 1967, p. 191)

---

[14]See also the following quote from Avicenna: "And the quiddities of things may be in individual things, and they may be in the mind; so they have three respects: the respect of quiddity inasmuch as it is that quiddity is not added to one of the two modes of existence, nor to what is attached to the quiddity, insofar as it is in this respect. And quiddity has a respect insofar as it is in individuals. And there accidents which make particular its existence in that are attached to it. And it has a respect insofar as it is in the mind. So there accidents that make particular its existence in that are attached to it; e.g., being a subject and being a predicate, and universality and particularity in predication...." (Avicenna 1952; quoted from Bäck 1996, pp. 133–134).

According to Owens, Aquinas took the first alternative, while Scotus took the second one. Indeed, Scotus heartily adopts the Avicennian thesis that a common nature is neither a universal nor an individual. He wrote:

> I say, then, according to Avicenna in Bk V of his *Metaphysics*, 'Equinity is just equinity; of itself it is neither one nor many, universal nor singular.' Consequently, just as it is not only singular or only universal in the mind, so also in the extramental world of nature it is neither one nor many of itself. Hence, nature of itself includes neither this nor that numerical unity. (Scotus 1992, p. 17; Dist. 3, q. 1, n. 30, 31)

Common nature itself, according to Scotus, is neither universal nor individual. It has less than numerical unity, though surely independent of any existence.

One interesting point to note is that, to a certain extent, Shapiro might concur with Scotus. Mathematical structures are independent of existence of mathematical systems or mathematical objects. They could be *ante rem* in that sense. I think, however, they do not have to be like Platonic Forms, which have numerical unity. In other words, I would like to know whether Shapiro's *ante rem* mathematical structrualism could be represented as a mathematical Scotism. Apparently it is meant to be a compromise between Plato's and Aristotle's positions. By highlighting the *ante rem* aspect of mathematical structures, Shapiro might want to emphasize the fact that his mathematical structuralism is a true successor of mathematical Platonism. But there is certainly a trade-off. Scotistic-Aristotelian reading of Shapiro may free Shapiro from all worries stemming from Benacerraff's challenge against mathematical Platonism without giving up the *ante rem* aspect of mathematical structures.[15]

It is not clear who to blame for the sudden disappearance of mathematical Aristotelianism around 1900. As one can witness from Quine's influential paper, mathematical Aristotelianism is simply gone from the list of rival theories in philosophy of mathematics (Quine 1948).[16] In order to sense that it could be troublesome, it would be enough to invoke the issue of the application of mathematics. Thanks to mathematical Aristotelianism, philosophers and mathematicians in the 17th, 18th, and 19th centuries did full justice to the problems of the application of

---

[15]Needless to say, what I write in this paragraph is at best highly speculative. Interestingly, one anonymous reviewer anticipated all this: "Scotus's notion of 'nature' is hardly applicable to abstract mathematical entities. But the difficulty is not only that 'nature' in Scotus's sense is tailored to something else than to abstract mathematical entities. As is known, the specific nature for Scotus and for the scholastic-Aristotelian tradition is qualitative, or better, quidditative, but such nature cannot be resolved into a bunch of properties! So it would be difficult to assess Scotus's position as relevant to the contemporary issue on haecceitas as a property". I would concede fully the difficulties involved in demonstrating the relevance of Scotus' theory of common nature to contemporary issues in mathematical structuralism. But wasn't the reviewer too pessimistic because of the implicitly assumed mathematical Platonism? As is well-known, Scotus and all other Aristotelians are ready to employ the theory of abstraction.

[16]Quine wrote: "The three main mediaeval points of view regarding universals are designated by historians as *realism*, *conceptualism*, and *nominalism*. Essentially these same three doctrines reappear in twentieth-century surveys of the philosophy of mathematics under the new names *logicism*, *intuitionism*, and *formalism*" (Quine 1948).

mathematics (see Park 2009). On the other hand, application of mathematics been unduly ignored by the twentieth century philosophers of mathematics. Of course, there have been some notable exceptions such as Steiner (1975, 1998), and now there seems to be a revival of interest in Aristotelianism in mathematics (Franklin 2014). In view of all this, it would be unfortunate, if Keränen and Shapiro are talking at cross purposes. Likewise, it would be regrettable, if philosophers of mathematics give up the dialogue with the history of ontology too early.

References

Adams, R. M. (1979). Primitive thisness and primitive identity. *Journal of philosophy, 76,* 5–26.

Andrews, R. (2010). Haecceity in the metaphysics of John Duns Scotus. In L. Honnefelder, et al. (Eds.), *Johannes Duns Scotus 1308–2008; Die philosophischen Perspektiven seines Werkes/ Investigations into his Philosophy.* Proceedings of "The Quadruple Congress" on John Duns Scotus, Part 3. Münster: Aschendorff Verlag, and Bonaventure: Franciscan Institute.

Angelelli, I. (1967). *Studies on Gottlob Frege and traditional philosophy.* Dordrecht: D. Reidel.

Bäck, A. (1996). The triplex status naturae and its justification. In I. Angelelli & M. Cerrezo (Eds.), *Studies in the history of logic.* Berlin: Walter de Gruyter.

Bergmann, G. (1964). *Logic and reality.* Madison: University of Wisconsin Press.

Bergmann, G. (1967). *Meaning and existence.* Madison/Milwaukee/London: University of Wisconsin Press.

Chisholm, R. M. (1986). Possibility without haecceity. In P. French, et al. (Eds.), *Midwest studies in philosophy XI: Studies in essentialism.* Minneapolis: University of Minnesota Press.

Cross, R. (1999). *Duns Scotus.* Oxford: Oxford University Press.

Cross, R. (2003, 2014). *Medieval theories of haecceity. In Stanford Encyclopedia of Philosophy.* http://plato.stanford.edu/entries/medieval-haecceity/.

Franklin, J. (2014). *Aristotelian realist philosophy of mathematics.* Palgrave MacMillan.

Gracia, J. J. E. (1996). Individuality and the individuating entity in Scotus's ordinatio: An ontological characterization. In L. Honnefelder, R. Wood & M. Dreyer (Eds.), *John Duns Scotus: Metaphysics and ethics* (pp. 229–249). Leiden: E.J. Brill.

Keränen, J. (2001). The identity problem for realist structuralism. *Philosophia Mathematica (III), 9,* 308–330.

Keränen, J. (2006). The identity problem for realist structuralism II: A reply to Shapiro. In MacBride (2006b, pp. 146–163).

Ladyman, J. (2005). Mathematical structuralism and the identity of indiscernibles. *Analysis, 65,* 218–221.

Lloyd, A. C. (1981). *Form and universal in Aristotle.* Liverpool: Francis Cairns.

Losonsky, M. (1987). Individual essence. *American Philosophical Quarterly, 24,* 253–260.

MacBride, F. (2005). Structuralism reconsidered. In S. Shapiro (Ed.), *Oxford handbook of philosophy of mathematics and logic* (pp. 563–589). Oxford: Oxford University Press.

MacBride, F. (2006a). What constitutes the numerical diversity of mathematical objects? *Analysis, 66,* 63–69.

MacBride, F. (Ed.). (2006b). *Identity and modality.* Oxford: Oxford University Press.

McMichael, A. (1983). A problem for actualism about possible worlds. *Philosophical Review, 92* (1), 49–66.

Owens, J. (1967). Common nature: A point of comparison between thomistic and scotistic metaphysics. In J. F. Ross (Ed.), *Inquiries in medieval philosophy* (pp. 185–209). Westport, Conneticut: Greenwood Pub. Co.

Park, W. (1988). *Haecceitas and the bare Particular: A study of Duns Scotus' theory of individuation* (Ph.D. dissertation). Buffalo: State University of New York at Buffalo.

Park, W. (1990). Haecceitas and the bare particular. *Review of Metaphysics, 44,* 375–398.

Park, W. (1998). Haecceitas and individual essence in Leibniz. In S. Brown (Ed.), *Meeting of the minds: The relationship between medieval and modern philosophy* (pp. 359–375). Belgium: Societe Internationale pour lÉtude de la Philosophie Medievale.

Park, W. (2000). Toward a scotistic modal metaphysics. *Modern Schoolman, 77,* 191–198.

Park, W. (2009). The status of scientiae mediae in the history of mathematics: Biancani's case. *Korean Journal of Logic, 12*(2), 141–170.

Park, W. (2016). What if haecceity is not a property? *Foundations of Science, 21,* 511–526.

Plantinga, A. (1978). The Boethian compromise. *American Philosophical Quarterly, 15*(2), 129–138.

Quine, W. V. O. (1948). On what there is. *Review of Metaphysics, 2*(5), 21–36.

Rosenkrantz, G. S. (1993). *Haecceity: An ontological essay.* Dordrecht: Kluwer.

Scotus, J. D. (1992). *Duns Scotus' early oxford lecture on individuation* [Latin Text and English Translation by A. B. Wolter, Santa Barbara]. CA: Old Mission Santa Barbara.

Shapiro, S. (1997). *Philosophy of mathematics: Structure and ontology.* NewYork: Oxford University Press.

Shapiro, S. (2006a). Structure and identity. In MacBride (2006b, pp. 109–145).

Shapiro, S. (2006b). The governance of identity. In MacBride (2006b, pp. 164–173).

Shapiro, S. (2008a). Reference to indiscernible objects. In M. Pelis (Ed.), *Thelogica Yearbook 2008* (pp. 223–235). London: College Publications.

Shapiro, S. (2008b). Identity, indiscernibility, and ante rem structuralism: The tale of i and $-i$. *Philosophia Mathematica (III), 16,* 285–309.

Shapiro, S. (2012). An "i" for an i: Singular terms, uniqueness, and reference. *The Review of Symbolic Logic, 5*(3), 380–415.

Steiner, M. (1975). *Mathematical knowledge.* Cornell: Cornell University Press.

Steiner, M. (1998). *The applicability of mathematics as a philosophical problem.* Cambridge, MA: Harvard University Press.

Ujvári, M. (2013). *The trope bundle theory of substance: Change, individuation and individual essence.* Frankfurt: OntosVerlag.

Wolter, A. B. (1992). Introduction. In Scotus (1992), pp. ix-xxvii.

Chapter 12
Biancani on *Scientiae Mediae*

Abstract We can witness the recent surge of interest in the controversy over the scientific status of mathematics among Jesuit Aristotelians around 1600. Following the lead of Wallace, Dear, and Mancosu, I propose to look into this controversy in more detail. For this purpose, I shall focus on Biancani's discussion of *scientiae mediae* in his dissertation on the nature of mathematics. From Dear's and Wallace's discussions, we can gather a relatively nice overview of the debate between those who championed the scientific status of mathematics and those who denied it. But it is one thing to fathom the general motivation of the disputation, quite another to appreciate the subtleties of dialectical strategies and tactics involved in it. It is exactly at this stage when we have to face some difficulties in understanding the point of Biancani's views on *scientiae mediae*. Though silent on the problem of *scientiae mediae*, Mancosu's discussions of the Jesuit Aristotelians' views on *potissima* demonstrations, mathematical explanations, and the problem of cause are of utmost importance in this regard, both historically and philosophically. I will carefully examine and criticize some of Mancosu's interpretations of Piccolomini's and Biancani's views in order to approach more closely what was really at stake in the controversy.

1 Introduction

We can witness the recent surge of interest in the controversy over the scientific status of mathematics among Jesuit Aristotelians around 1600. Both historians and philosophers of science (and mathematics) have good reason to be enthusiastic about scrutinizing this controversy. For, it must not only deepen our understanding of the role of mathematics in the scientific revolution, but also present a nice case study for Aristotelian scientific methodology and philosophy of mathematics in action. Following the lead of Wallace, Dear, and Mancosu, I propose to look into

---

This chapter was originally published as Park (2009). An earlier version was read at the international conference on "The Classical Model of Science" held at Amsterdam in 2007.

© Springer International Publishing AG, part of Springer Nature 2018
W. Park, *Philosophy's Loss of Logic to Mathematics*, Studies in Applied Philosophy, Epistemology and Rational Ethics 43, https://doi.org/10.1007/978-3-319-95147-8_12

this controversy in more detail. For this purpose, I shall focus on Biancani's discussion of *scientiae mediae* (subordinate, subalternated, intermediate, middle, or mixed sciences; applied mathematics or mixed mathematics) in his dissertation on the nature of mathematics. In the discussion of the perennial problem of unity or disunity of sciences, Aristotle's general prohibition of *metabasis* has always played a key role by securing the autonomy of individual sciences. But some disciplines like astronomy or music provide us with obvious counterexamples to this rule by applying the results of pure mathematics. According to Dear (1995), Aristotle's ad hoc solution of this problem by classifying them as *scientiae mediae* caused a battle for later Jesuit Aristotelians on the problem of scientific status of *scientiae mediae*, and thereby the problem of relationship among pure mathematics, applied mathematics, and physics. From Dear's and Wallace's discussions, we can gather a relatively nice overview of the debate between those who championed the scientific status of mathematics and those who denied it. But it is one thing to fathom the general motivation of the disputation, quite another to appreciate the subtleties of dialectical strategies and tactics involved in it. It is exactly at this stage when we have to face some difficulties in understanding the point of Biancani's views on *scientiae mediae*. It is in part due to the strange structure of his dissertation, which concludes with surprisingly brief discussion of *scientiae mediae*. More seriously, some of Biancani's views seem to be conflicting each other. For example, it is not clear why he emphasized the difference in kind between pure and applied mathematics if he would have *scientiae mediae* in the realm of the ideal model of science. Though silent on the problem of *scientiae mediae*, Mancosu's discussions of the Jesuit Aristotelians' views on *potissima* demonstrations, mathematical explanations, and the problem of cause are of utmost importance in this regard, both historically and philosophically. I will carefully examine and criticize some of Mancosu's interpretations of Piccolomini's and Biancani's views in order to approach more closely what was really at stake in the controversy.

2 Wallace and Dear on the Motivations of Jesuit Debates on *Scientiae Mediae*

In both Dear and Wallace we can find very useful perspective for understanding Biancani in more general context of the period around 1600. For they tried to understand Biancani against the background of their detailed discussion of Clavius, the reformer of mathematical education among the Jesuits that promoted scientific knowledge in the period of Scientific Revolution.

Wallace starts his discussion of Biancani immediately after the following remark: "The full effect of Clavius's program, was not seen until students whom he prepared himself under the new Ratio Studiorum completed their studies and began to write on the nature of the mathematical disciplines." (Wallace 1984, p. 141). Dear also points out that even though the Jesuits uplifted the status of mathematics and the results were reflected in the educational program, the exact relation between mathematics and philosophy was a matter of controversy. And he pins down Clavius's role as a defender of the philosophical status of mathematics against the opponents in the controversy around the curriculum in Rome in 1580s. He also made it explicit that Clavius was not attacking the straw men but the real objectors like Piccolomini, Pereyra, and Coimbra commentators. In Dear's discussion we can also find many suggestive comments on the strategies and tactics of Clavius in the controversy. For example, according to Dear, Clavius suggested to let mathematicians attend the regular disputation as well as philosophers. Clavius argued for the necessity of mathematics in order to study natural philosophy by using the point that mathematical astronomy is needed for cosmological studies (Dear 1995, pp. 35–36).

Even more interesting is the fact that, according to Dear, arguments given by Clavius for his claim were merely arguments from authorities. By resorting to Aristotle's authority, for example, Clavius emphasized the fact that mathematics, physics, and metaphysics had been the three constituents of speculative philosophy from the antiquity. Also he claimed the excellence of mathematics in terms of Ptolemy's authority. Moreover, according to Dear, Clavius evaded the crucial objections that mathematical disciplines are not scientific. In other words, Clavius's strategy in the controversy was not launching any positive argument but simply subsuming mathematical sciences under the Aristotelian model of ideal science (Dear 1995, pp. 37–39).

In order to appreciate the import of Dear's comments we need to grant a fair hearing to his explanation of how a crucial problem was raised in Aristotelian methodology, which was destined to bring about the controversy among the Jesuit Aristotelians before and after 1600. According to Dear's description, individual sciences in Aristotle's model must be based on unique and proper principles that would function as the major premise in syllogistic demonstrations. As a consequence, individual sciences are separated sharply from each other. And that is a logical necessity expressed in the methodological principle of homogeneity. That homogeneity requires that the principle of an individual science must be related to the genus as its object in order to secure deductive link. But some disciplines such as astronomy and music clearly violate this rule by applying the results of pure mathematics to celestial motions and sounds. Aristotle made special arrangement for these sciences by classifying them as subordinated to higher scientific disciplines. According to Dear, such an Aristotelian solution was more or less an ad hoc device for the problem of classifying sciences. And, as a consequence, it was destined to bring about the controversy among the Jesuit Aristotelians as to "whether demonstrations in a subject such as optics yielded true scientific knowledge if the presupposed theorems of geometry were not proved at the same time" (Dear 1995, pp. 38–39; Here Dear refers to Wallace 1984, p. 134).

So far, Dear's description of the problem situation is quite persuasive and illuminating. But, much more than that, Dear presents a subtle and ingenious interpretation of Clavius's strategy in the situation. According to Dear, since the attempt itself to assimilate applied disciplines to the general model of science made it clear that Aristotle granted the scientific status to all mathematical disciplines, such an Aristotelian approach was in full service for Clavius's purposes. Finally, Dear's ingenious interpretation arrives at the pinnacle in his claim that unlike Clavius, who was able to claim the scientific status of mathematics without involvement with the thorny problem of cause, Biancani directly tackled the problem of cause (Dear 1995, p. 40).

3   The Chapter on *Scientiae Mediae*

Happily beginning to understand the general motivation of the disputation, one might desire to appreciate the subtleties of dialectical strategies and tactics involved in it. For this reason, it is truly timely that Mancosu published Klima's English translation of Biancani's dissertation on the nature of mathematics as an appendix to his book *Philosophy of Mathematics and Mathematical Practice in the Seventeenth Century* (Mancosu 1996). Furthermore, Mancosu discussed Biancani's philosophical views on mathematical sciences rather extensively in several writings (Mancosu 1991, 1992, 1996).

It is exactly at this stage when we have to face some difficulties in understanding the point of Biancani's views on *scientiae mediae*. It is in part due to the strange structure of his dissertation, which concludes with surprisingly brief discussion of *scientiae mediae*. Thanks to Mancosu and Klima, now more people have come to know that Biancani's *De Mathematicarum Natura Dissertatio* (1615a) was published as an appendix to his *Aristotelis Loca Mathematica*, which was an excerpt of passages on mathematics in Aristotelian corpus. Biancani's dissertation on the nature of mathematics consists of five chapters. The first chapter deals with the subject of mathematics. Middle term in geometrical demonstration is discussed in the second chapter, where the crucial issue is whether geometrical demonstrations are perfect (*potissima*). In chapter three he criticizes opponents' errors. The fourth chapter is an attempt to argue for the excellence of mathematics. At the end of the chapter four Biancani enumerates all the conclusions he secured from his discussion of pure mathematics in the first four chapters. And, finally in chapter five, he discusses *scientiae mediae* such as astronomy, optics, mechanics, and music. Why does he discuss all these themes in this particular order? In particular, why does he discuss *scientiae mediae* only at the end? Is it due to their prime importance, or their relative unimportance? Since Biancani's dissertation on the nature of mathematics is itself an appendix to his *Aristotelis Loca Mathematica*, he might be excused for his all too frequent cross-references in the former to the latter. What is interesting is the fact that his final chapter follows the same pattern of referring back to his previous discussions. Be that as it may, his chapter on *scientiae mediae* is all too brief to allow sound understanding of his views about them.

Roughly speaking, what we can read in this chapter on *scientiae mediae* is this. In the opening and the closing paragraph, we can confirm his conclusion that *scientiae mediae* present us with perfect demonstrations. He resorts to Aristotle's authority to show that *scientiae mediae* do have demonstrations of causes: "For here the 'what' is to be known by those who perceive, but the 'why' by the mathematicians, for they have the demonstrations of the causes." (M 206, B 313; Anal. Post. 79a 3–6). Then, he provides us with a few examples of alleged perfect demonstrations in applied mathematical sciences. What is troublesome is that his discussion of them is too succinct, if not cryptic. Our expectation to find some further hints as to the implications of his views on *scientia mediae* is almost completely betrayed.

Let me quote his discussion of some examples: one from astronomy and another from optics.

> To begin with astronomy, isn't the demonstration of the eclipse of the moon (even by the testimony of Aristotle and his commentators, especially Zabarella) a perfect demonstration? For it renders evident the proper and adequate cause of the property [*affectio*] in question, i.e., the eclipse, namely, the interposition of the Earth. But we should say the same of the solar eclipse, the cause of which is shown to be the interposition of the Moon. And that these demonstrations were discovered by the astronomers is known from their books, as well as the fact that they use geometrical media, namely, circle, diameter, and diametric opposition, and that therefore how certain they are is obvious from the infallible prediction of eclipses." (M. 206, B. 314)

> Again, in optics we do not lack perfect demonstrations either. For example, why is the eye spherical? So that perpendicular lines can fall on it from every direction. But why perpendicular lines? In order to produce distinct sight. Here you have the final cause. (M. 207, B. 314)

Could we understand what a perfect demonstration should be like from such brief discussions as these? After all, wherein lies the novelty of Biancani's own contribution?

Laird's assessment that Biancani defended mathematics more with rhetoric and examples than with detailed exposition of Aristotle seems not entirely unfair. Also, even after having considered what Biancani did in *Aristotelis loca mathematica*, he points out quite justifiably that Biancani failed to solve the difficulties raised in the commentaries on the *Posterior Analytics* (Laird 1997, p. 266).

4   Biancani's Target

But we have to note that there is one potentially revealing remark at the closing part of chapter five. For, there he claims:

> From these it manifestly appears that mathematical sciences have perfect demonstrations, whose causes are so distinct from their effects that no calumnies can do any harm to them. Therefore, even if our opponent could prove, which they never can, that geometry and arithmetic lacks them, they would have to admit this concerning the other [disciplines] mentioned above that they reason by all genera of causes, and that they excel with such clarity that they leave nothing in ambiguity or controversy. (M. 208, B. 315)

This exceptional remark is interesting in view of the fact that for the last three decades Quine-Putnam indispensability argument has been counted as the only persuasive argument for realism in philosophy of mathematics (see Colyvan 2001, Maddy 1997, Putnam 1979, and Quine 1980). For, in both Biancani and twentieth century realists, applied mathematics seems to play the role of the last bulwark for realism in mathematics. Be that as it may, Biancani's remark seems to give us a definite hint for identifying the prime target of his dissertation.

In chapter three of his dissertation, Biancani criticizes seventeen errors of his opponents. Fortunately, at least some of them have definite relevance for *scientiae mediae*: i.e., the third, the fourteenth, and the sixteenth. The third errors stems from Plato's remark in Book 7 of the *Republic* that mathematicians dream about quantity. The fourteenth error has something to do with the alleged ignobility of the subject matter of mathematics. And, the sixteenth error is related to an inconsistency involved in the opponents of Biancani. For my present purpose, what does matter is the sixteenth error:

> The sixteenth is [that] which they put forward by asking [first] in general whether mathematics has perfect demonstrations, then later in the discussion they bring up several points against mathematics, and at the end of the treatise they claim that these concern only geometry and arithmetic. Wherefore, unless the reader peruses everything to the end, which rarely happens, he will be deceived, for he will think that all mathematical sciences were concerned, while the authors themselves acknowledge that they have never spoken about applied mathematics, i.e., astronomy, music, optics, and mechanics, which they readily admit to be true demonstrative sciences. (M 203, B. 310)

Together with Biancani's revealing remark discussed above, this point seems to indicate how important it is for him that his opponents grant the status of demonstrative science to *scientiae mediae*. For, now it becomes clear that we have to sharply distinguish between (1) those adversaries of mathematics who concedes the existence of perfect demonstrations in pure mathematics but denying it for *scientia mediae*, (2) those who concede it for *scientiae mediae* but denying it for pure mathematics, and (3) those who deny it for both pure mathematics and *scientiae mediae*. It is obvious that he has to respond somewhat differently for strategic purposes depending upon the different type of his opponents. As a consequence, now we can see that the prime target of Biancani's dissertation must be of the second type, i.e., those who concede the existence of perfect demonstrations in *scientiae mediae* but denying it for pure mathematics.

5 Biancani on Perfect Demonstrations

In chapter three of his dissertation, Biancani starts his discussion by his claim that the mathematicians of his age are "compelled to guard by every effort what was so far their safe, ancient, and rightful possession from some recent thinkers who strive to take it away" (B294, M184). And, in the next sentence, he raises a rhetorical question as to whether there "was ever a philosopher of stature before Alessandro

Piccolomini who attempted to rob geometers of perfect demonstrations". By exploiting Piccolomini's claim for his own originality, Biancani declares assuredly that there has been none. Biancani's portrayal of the situation is apparently confirmed by Mancosu. According to Mancosu, in his work *Commentarium de Certitudine Mathematicarum Disciplinarium* (1547),

> Piccolini challenged the traditional argument that mathematical sciences possess the highest degree of certainty because they make use of the highest type of demonstration, the *potissima* demonstration, defined by him as that which gives at once the cause and the effect (*simul et quia et propter quid*). (Mancosu 1996, p. 12)

On the other hand, in Laird's rather extensive comparisons and assessments of medieval and renaissance philosophers' views on the demonstrative power of mathematical sciences, we seem to find an entirely different picture. According to him, most of them except for Aquinas and Zabarella, denied the existence of *potissima* proofs in *scientiae mediae*.[1] Furthermore, according to Laird, despite their citations of Aquinas and Zabarella, none of the Jesuit commentators associated with the Collegio Romano allowed *propter quid* demonstrations for mixed sciences (Laird 1997, p. 260). Finally, according to Laird, Biancani was a rare exception to the general tradition for his "rare confidence in the demonstrative power of the mixed sciences" (Laird 1997, p. 266).

Of course, we must be able to resolve many apparent inconsistencies involved in these reports by sharply distinguishing between the problem of *potissima* proof in pure mathematics and that of *scientiae mediae*, as Biancani did. Still we need to be alert to some potentially misguiding emphases and perspectives as well as incompatible reports. For, even if we restrict our concern to the problem of *potissima* proof in *scientiae mediae*, we find a contradiction between Biancani's and Laird's perceptions of the situation. If Biancani is right, then there was no disagreement about the existence of *potissima* proofs in *scientiae mediae*. On the other hand, if Laird is right, then there was even more serious disagreement about the existence of *potissima* proofs in *scientiae mediae* than in pure mathematics. Furthermore, it is not a minor problem whether to appreciate Biancani as championing the majorities or minorities in the disputes.

It seems to me that there is room for suspicion whether Laird and Mancosu are a bit unfair to Biancani in their description of the controversy about the scientific status of mathematical sciences. Above all, both tend to call or characterize the

---

[1]"In accordance with their understanding of Aristotle, the commentators all sought to keep the terms of premises and conclusions within the same subject genus. For most of them this meant that a mathematical proof in a mixed science is quia, not propter quid, since it is not made through the proximate, necessarily physical cause of the composite predicate's adhering in the composite subject. And when a mathematical middle term proved a mathematical predicate of a physical subject, the proof was usually considered quia through the remote cause. Only Zabarella and, in a more limited way, Aquinas, allowed for propter quid mathematical demonstrations in the mixed sciences." (Laird 1997, pp. 259–260).

controversy in terms of the certainty of mathematical sciences. As we saw above, Macosu presents Piccolomini as challenging the traditional argument that mathematical sciences possess the highest degree of certainty because they use *potissima* demonstrations. Laird too, again as we saw above, testifies "the widespread doubt over the certainty of mathematics in general expressed by several prominent Jesuit professors at the Collegio Romano." (Laird 1997, p. 260). Mancosu's evidence for the existence of the allegedly traditional argument is Piccolomini's list of authorities including Aristotle, Averroes, Albert, Aquinas, and Nifo (Mancosu 1992, 244; See also M 187, B.297). In order to prove the existence of a tradition, however, it may not be good enough to enumerate a list of authorities. Nor do I find in Biancani's dissertation any emphasis on the certainty of mathematical sciences.

If we assume Mancosu's perspective, it appears that the alleged traditional argument that mathematical sciences possess the highest degree of certainty might be another expression of the Platonic position according to which mathematics is superior to physics. According to the more widely accepted view in history of science, however, possibly more Aristotelian position according to which physics is superior to mathematics was more dominant in both medieval and renaissance period. For example, Westman wrote: "If, however, an astronomer were determined to reconcile physical and mathematical issues, it would be customary within the Aristotelian tradition (which prevailed within the universities) to defer to the physicist, for in the generally accepted medieval hierarchy of the sciences, physics or natural philosophy was superior to mathematics." (Westman 1986, p. 78). In the footnote, he identifies that tradition as going back to Albertus Magnus, and contrasts it with the opposing tradition at Oxford (Westman 1986, p. 105). In view of the existence of these two rival traditions, I think, Mancosu's presentation of Piccolomini as challenging the only extant tradition could be a distortion of history.

The reason why Piccolomini grants the certainty of mathematics might be understood as a subtle strategy for not allowing *potissima* proof for pure mathematics. According to Mancosu's exposition of Piccolomini's position, Piccolomini thinks that mathematics possesses the highest degree of certainty because mathematical objects are created by human mind (Mancosu 1992, 244). Since such a position must deny the existence of mathematical objects independent of human mind, mathematical proofs apparently cannot be the *potissima* proof as Piccolomini defines. In other words, it could be a trap to formulate the issue in terms of the certainty of mathematics, and Biancani was wise enough to avoid the issue of mathematical certainty itself.

Laird provides us with a hint as to why some Jesuit commentators wanted to deny the certainty of mathematical sciences. For he writes: "they often doubted whether demonstrations within the mixed sciences could even be demonstratively certain, let alone *propter quid*." (Laird 1997, p. 260). Citing Toledo and Pereyra as his examples, Laird goes on to point out that "demonstrations in physics answer to Aristotle's ideal of the most powerful demonstrations." (Laird 1997, p. 261). One interesting consequence of such a view for *scientiae mediae*, according to Laird, is that "they have all the disadvantages of being mathematical-they do not demonstrate through cause-and all the uncertainty of physics-their subjects involve physical things." (Laird 1997, p. 261).

Now we can see that all the possible types of denying *potissima* proof to mathematical sciences indeed have actual instances in history. First, in Pereyra, we have the most radical position of denying *potissima* proof in both pure mathematics and in *scientiae mediae*. Secondly, as was pointed out, Biancani's primary targets were those who denied *potissima* proofs for pure mathematics but allowing them in *scientiae mediae*. Finally, more common opinion must have been allowing *potissima* proofs for pure mathematics but denying them for *scientiae mediae*. Since Biancani grants *potissima* proofs for both pure and *scientiae mediae*, it is curious why he does not bother with the first and the third type of adversaries in his dissertation.

Biancani explicitly mentions Pereyra as one of the two who are following Piccolomini's footsteps (M 187, B. 297). Also, there are passages indicating that Biancani takes Pereyra's view quite seriously. For example, we may cite Biancani's interpretation of 32d proposition of the first book of Euclid's *Elements* as evidence. Mancosu discussed how Pereyra raised the issue of causality by using the same proposition as a clear counterexample to the causal theory of mathematics. According to Mancosu's exposition, Pereyra's criticism is that (1) the middle term in the proof of the 32d proposition is the appeal to auxiliary segments and to the external angles, but (2) the auxiliary segments and the external angles cannot be the true formal cause of the equality (Mancosu 1996, pp. 14–15). Interestingly, Biancani uses the same proposition from Euclid's *Elements* as an example of the material cause in pure mathematics. Also, he notes that "the parallel line by which the [external] angle is divided is drawn in order to find the medium of the demonstration, but it is by no means the medium itself." (M191–192, B 300); See also Piccolomini, p. 97) If so, it needs some explanation why Biancani simply ignores Pereyra's denial *of potissima* proof both for pure mathematics and *scientiae mediae*.

Even more strange is that Biancani does not discuss in detail the relationship between the subalternating science and the subalternated science. As in the discussion of Wallace and Dear, when we first hear about Aristotle's prohibition of *metabasis* and his ad hoc solution of allowing subalternated sciences, we tend to think that in subalternated science we can merely have *quia* demonstrations and that only in subalternating science we can have *propter quid* explanations. Indeed, we get the same impression from the 13th chapter of the first book of Aristotle's *Posteior Analytics*. According to recent studies on *scientia mediae* in medieval and renaissance commentaries on Aristotle's *Posterior Analytics*, however, there were several ingenious interpretations regarding the relationship between demonstrations in subalternating science and subalternated science (Laird 1997, 1983; Livesey 1982, 1989).

Laird's discussion of Robert Grosseteste's case can be a nice sample for showing what kind of issues are involved (Laird 1983, p. 36f). Suppose that we have a proposition to be proved in optics:

> that every two angles of which one constitutes the ray incident upon a mirror and the other the reflected ray are two equal radiant angles.

According to Grosseteste, we need to appropriate a proposition from geometry in order to prove it. In this case, the proposition from geometry is this:

> Of any pair of triangles of which one angle of one is equal to one angle of the other and the sides containing the equal angles are proportional, the corresponding remaining angles are equal.

Grosseteste appropriates this geometrical proposition to optics by adding the condition "radiant" to its terms:

> of any two radiant triangles of which one radiant angle of one is equal to one radiant angle of the other and the radiant sides containing the equal radiant angles are proportional, the corresponding remaining radiant angles are equal.

Ultimately, as Laird represents the situation, we will have a syllogism of the following pattern:

$$\langle rA \rangle \quad \begin{array}{l} rS \ rM \\ rM \ rP \\ rS \ rP \end{array}$$

Within geometry the purely geometrical proposition MP could be used in a demonstration as follows:

$$\langle A \rangle \quad \begin{array}{l} S\,M \\ M\,P \\ S\,P \end{array}$$

Now, according to Laird, Grosseteste counts $\langle A \rangle$ as a demonstration *propter quid*, while $\langle rA \rangle$ as a demonstration *quia*: "For the cause of the equality of radiant angles is not in the major premise borrowed from geometry and appropriated to *perspectiva*, but it is rather in the nature of light and the regularity of nature" (Laird 1983, p. 40; cf. Robert Grosseteste, *Comm. Post. Anal.* I. 8, 93–101).

Though interesting, I believe that Grosseteste might be misrepresenting Aristotle's intention here. But I would not discuss this matter any further. For my present purpose, what is needed is to note simply that Latin commentators after Robert Grosseteste had to take stance in this matter on one way or another. For, as Laird points out, Grosseteste's commentary was not only the first systematic exposition of Aristotle's *Posterior Analytics* in the Latin West but also "the starting point for all subsequent discussions of the logical problems of demonstration within the intermediate sciences" (Laird 1983, p. 53). Further, insofar as the problem of the existence of *potissima* demonstrations in *scientiae mediae* does matter, the key lies in answering whether ⟨rA⟩ presents us a *propter quid* demonstration.

Apparently, Biancani does not discuss this issue extensively. But again he leaves us a revealing remark in an intriguing context:

> Finally, in the fourth place, we confirm the same point by the common authority of all ancient authors, who always call geometrical proofs by appropriation [*per antonomasiam*], and not reasons, or opinions, or tenets [*sententiae*], as it happens in other parts of philosophy. But let us turn from authority to reason." (M. 188, B. 298)

In chapter two of his dissertation on the nature of mathematics, where he discusses *potissima* demonstrations, Biancani provides us with four arguments based on authority and three arguments based on reason. In his arguments based on authorities, Binacani resorts to Aristotle, Plato, and Proclus, and devotes large space for each of them. Unlike these authorities, Biancani writes just one sentence for the fourth argument based on authority, as we have just seen. Why? Nor does he identify the authoritative philosopher either. Why? Even though the key word, i.e., "appropriation" indicates the opportunity to discuss Grosseteste, Biancani is referring to "the common authority of all ancient authors". Why?

6 Biancani's Strategy

Mancosu is almost silent on the problem of *scientiae mediae* in his book, for he declares that he will not discuss the relationship between mathematic and physics (Macosu 1996, p. 3, p. 213). In fact, even when confined to pure mathematics, Mancosu's discussions of the Renaissance Jesuit Aristotelians' views on *potissima* demonstrations, mathematical explanations, and the problem of cause are interesting and of utmost importance both historically and philosophically. Further, Mancosu does not investigate the problem of mathematical explanation merely as of historical curiosity but as of utmost philosophical value (Mancosu 2000, 2001). I believe that Mancosu could have been indebted greatly to Biancani's dissertation in this regard, for the problem of explanation has been rarely discussed in contemporary philosophy of mathematics.[2] From this point of view, the first chapter of Biancani's dissertation on the nature of mathematics, which deals with the problem of the subject matter of pure mathematics, the problem of mathematical abstraction, and the problem of definition in mathematics, seems to be a treasure house for fascinating issues for philosophers of mathematics.

Though tempting, it must be beyond the scope of this article to delve into some such fundamental philosophical issues in pure mathematics. Let me just draw your attention to one point that has definite relevance for understanding Biancani's strategy to handle his adversaries. Biancani begins his discussion of the first chapter

---

[2]Mark Steiner must be a notable exception in this regard. See Steiner (1975, 1998).

with a revealing claim that pure mathematics differs in kind from applied mathematics. More interestingly, he writes:

> Quantity abstracted from sensible matter is usually considered in two ways. For it is considered by the natural scientist and the metaphysician in itself, that is absolutely, insofar as it is quantity, whether it is delimited [terminata] or not; and in this way its properties are divisibility, locatability, figurability, etc. But the geometer and the arithmetician consider [quantity] not absolutely, but insofar as it is delimited, as are the finite straight or curved lines in continuous quantity... (B 289, M179)

What I find interesting and revealing here might be expressed by raising the following two questions: (1) Why does Biancani contrast pure mathematics with natural science and metaphysics, when we expect to hear about exactly wherein lies the difference between pure and applied mathematics?; (2) Why does he not compare mathematics, natural science, and metaphysics,[3] thereby highlighting the peculiarity of each, but comparing mathematics with "natural science and metaphysics lumped together"? As for the first question, the only possible explanation seems to this. Bicancani thinks that the difference in kind between pure and applied mathematics cannot be understood without contrasting pure mathematics with "natural science and metaphysics". As for the second question, we might speculate that Biancani is worried about the possibility of losing sight of the differentiating characteristic of mathematics from other speculative sciences (i.e., natural science and metaphysics) in case he adopts the traditional way of comparing the three speculative sciences. But what is the ultimate reason for his ways of thinking and worries? I cannot figure out any other than strategic, rhetorical, and political reasons.

The reason why Biancani depends on ancient and more recent authorities, thereby being parsimonious to cite medieval sources might be explained if we incorporate the general tendency of Renaissance scholars to recover the purity of ancient philosophers eliminating all Arabic and Latin medieval accretions (cf. Lohr 1999). Also, the reason why Biancani simply ignores Pereyra's radical denial of potissima proof both for pure and applied mathematics might be explained if we remember that Biancani, as a contemporary of Galileo. In other words, Biancani was in a position to exploit fully the external support from the remarkable success of applied mathematical sciences of his time.

But why does Biancani emphasize the difference in kind between pure and applied mathematics? If Dear is right in his interpretation, Biancani as well as

---

[3]Clavius provides us with a clear example of this way of comparing the three disciplines: "Because the mathematical disciplines discuss things that are considered apart from any sensible matter—although they are themselves immersed in matter—it is evident that they hold a place intermediate between metaphysics and natural sciences, if we consider their subjects, as is rightly shown by Proclus. For the subject of metaphysics is separated from all matter, both in the thing and in reason; the subject of physics is in truth conjoined to sensible matter, both in the thing and in reason; when, since the subject of the mathematical disciplines is considered free from all matter-although it[i.e., matter] is found in the thing itself-clearly it is established intermediate between the other two." (Clavius, "In disciplinas mathematicas prolegomena" in Opera mathematica, Vol. 1, p. 5; requited from Dear 1995, p. 37).

Clavius would include mathematical sciences within the realm of the Aristotelian ideal model of science. But, if so, would they include (1) pure mathematics only in that realm, or (2) applied mathematics only, or (3) both pure and applied mathematics? (1) cannot be the case, for that seems contradicting the overall project of securing the scientific status of applied mathematics. Nor (2) can be the case, for that would make Biancani's entire discussion of the excellence of pure mathematics futile and useless. If (3) is Biancani's intention, why does he emphasize the difference in kind between pure and applied mathematics?

7   Concluding Remarks

Following the lead of Wallace, Dear, and Mancosu, I tried to improve our understanding of the controversy among Renaissance Jesuit Aristotelians around the scientific status of mathematical sciences. Largely due to rhetorical nature of Biancani's work on the nature of mathematics, however, we are left with more puzzles and problems. At best, we might fathom Biancani's hidden agenda as follows:

> If we show that applied mathematics belongs to the same camp with physics and metaphysics, and if we loosen the alertness of our adversaries by emphasizing the difference in kind between pure and applied mathematics, (since applied mathematics is nothing but the result of applying pure mathematics after all) they themselves might have accepted inadvertently the point that the entirety of physics (and possibly even metaphysics) is the application of pure mathematics.

Now, in view of all this, what do we have to learn from Biancani's discussion of *scientiae mediae*? I think that there are at least two lessons if we compare our current situation with that of Biancani. One has to do with the practice of philosophy of mathematics, the other with redrawing the map of human knowledge.

Philosophy of mathematics in the twentieth century largely ignored the problems of applying mathematics. Only quite recently, we began to realize the significance of the philosophical issues involved in the application of mathematics in all the different individual sciences. On the other hand, virtually all mathematicians from Biancani's time to 1900 had to struggle with the philosophical problems of applying mathematics. That means, we cannot afford to ignore Aristotelian philosophies of mathematics in 17th, 18th, and 19th centuries as antecedent cases in philosophy of applied mathematics.

In the age of ever growing science, engineering, and technology, philosophers have no voice in organizing human knowledge. On the other hand, in the medieval and renaissance Latin West, philosophers were not only equipped with a general model of Aristotelian science but also able to detail it by probing questions as to the relationships among individual sciences. As was the case in Aristotle's discussion of geometry, optics, and the study of rainbow, the issue of subordination between mathematical sciences tends to be extended to other scientific disciplines as well.

As was hinted at above, some such debates are found in contesting the scientific status of theology in the middle ages. We also noted that, according to some Renaissance Jesuit Aristotelians, physics provides us with the best examples of perfect demonstrations. No matter what variations, complications, and changes were there, however, the dream of drafting the blueprint for the entire edifice of human knowledge modeling after the relationship between pure and applied mathematics was always there.

Biancani was so confident about the success of *scientiae mediae* as to ignore adversaries who denied *potissima* proofs for them. Mancosu was so sure about the promise of his project of rewriting the history of philosophy of mathematics as to be silent about the relation between mathematics and physics. As for the *fortuna* of philosophy and science in the 21st century, could we happily adopt their strategy?

References

Biancani, G. (Blancanus, Josephus). (1615a). *De Mathematicarum Natura Dissertatio*. Bologna. Available at http://archimedes.mpiwg-berlin.mpg.de/cgi-bin/toc/toc.cgi?dir-bian.

Colyvan, M. (2001). *The indispensability of mathematics*. New York: Oxford University Press.

Dear, P. (1995). *Discipline and experience: The mathematical way in the scientific revolution*. Chicago: The University of Chicago Press.

Laird, W. R. (1983). *The Scientiae Mediae in Medieval Commentaries on Aristotle's posterior analytics*. Ph.D. Dissertation, University of Toronto.

Laird, W. R. (1997). Galileo and the mixed sciences. In D. A. Di Lisca, et al. (Eds.), *Method and order in renaissance philosophy of nature* (pp. 253–270). Aldershot: Ashgate.

Livesey, S. J. (1982). *Metabasis: the interrelationship of the sciences in antiquity and the middle ages*. Ph.D. Dissertation, UCLA.

Livesey, S. J. (1989). *Theology and science in the fourteenth century*. Leiden: E.J. Brill.

Lohr, C. H. (1999). Aristotelian theories of science in the renaissance. *Sciences et religions: de Copernic a Galilee, 1540–1610*, 17–29.

Maddy, P. (1997). *Naturalism in mathematics*. Oxford: Oxford University Press.

Mancosu, P. (1991). On the status of proofs by contradiction in the seventeenth century. *Synthese, 88*, 15–41.

Mancosu, P. (1992). Aristotelian logic and Euclidean mathematics: Seventeenth century developments of the Quaestio de certitudine mathematicarum. *Studies in History and Philosophy of Science, 23*, 241–265.

Mancosu, P. (1996). *Philosophy of mathematics and mathematical practice in the seventeenth century*. Oxford: Oxford University Press.

Mancosu, P. (2000). On mathematical explanation. In *Grosholz* (pp. 103–119).

Mancosu, P. (2001). Mathematical explanation: Problems and prospects. *Topoi, 20*, 97–117.

Park, W. (2009). The status of Scientiae Mediae in the history of mathematics: Biancani's case. *Korean Journal of Logic, 12*(2), 141–170.

Piccolomini, A. (1547). Commentarium de Certitudine Mathematicarum Disciplinarum, Romae. Available at http://nausikaa2.mpiwg-berlin.mpg.de/digitallibrary/sevlet/Scaler?pn.

Putnam, H. (1979). *Mathematics, matter and method: Philosophical papers* (Vol. 1, 2nd ed.). Cambridge: Cambridge University Press.

Quine, W. V. O. (1980). *From a logical point of view* (2nd ed.). Cambridge: Harvard University Press.

Steiner, M. (1975). *Mathematical knowledge*. Ithaca, NY: Cornell University Press.

Steiner, M. (1998). *The applicability of mathematics as a philosophical problem*. Cambridge, MA: Harvard University Press.

Wallace, William A. (1984). *Galileo and his sources: The heritage of the Collegio Romano in Galileo's science*. Princeton: Princeton University Press.

Westman, R. S. (1986). The Copernicans and the churches. In D. C. Lindberg, R. N. Numbers (Eds.), *God and nature* (pp. 76–113). Berkeley: University of California Press.

Epilogue

We started our journey to the history of modern logic by raising a simple question as to the truly revolutionary character of Frege's new logic. Just as destined to any Socratic query, our inquiry was led to so many intricate questions relevant to the initial question. I am quite convinced that our efforts were not futile in that now we are able to pose more informed and penetrating questions that may shed light on the history of logic. Here I would like to highlight some of those questions that seem to me more urgent. This would allow me not only to indicate wherein lie the limitations of this monograph, but also where I find the clues for solving some of the toughest open problems in history and philosophy of logic.

It seems rather obvious from Part 1 and 2 that we cannot afford to ignore or postpone the resolution of Frege-Hilbert controversy any longer. It was very prudent for Bernays to conclude that the controversy was not over (Bernays 1942). As anyone can witness, however, such a prudent attitude of mathematicians has resulted in unbearable confusion in academia as well as in general public. Richard Robinson's book on definition may be still a recommendable point of departure for understanding definitions (Robinson 1963). For, it seems that our understanding of the problem of definition in mathematics has not made substantial progress. And, insofar as we fail to have a clear understanding of the role and function of definitions in mathematics, it is hard to expect any reasonable understanding of definitions in any other scientific disciplines or in our daily lives. But definitions are everywhere. What are the rules that we should follow in using definitions? How could we expect better communication in scientific or mundane matters, if we do not have any consensus on what definitions are?

Against such a background, we may welcome the recent trend among neo-Fregeans to discuss the problems of implicit definition in connection with the Fregean abstraction (Hale and Wright 2001; Cook 2007; Ebert and Rossberg 2016). It is still unclear what implications this trend might have to Frege-Hilbert controversy. Nevertheless, it seems evident that we will at least be able to learn a great deal about Frege's intellectual development in understanding the problems of definitions in mathematics. Certainly, we may expect to revive our interest in the

© Springer International Publishing AG, part of Springer Nature 2018   223
W. Park, *Philosophy's Loss of Logic to Mathematics*, Studies in Applied Philosophy,
Epistemology and Rational Ethics 43, https://doi.org/10.1007/978-3-319-95147-8

role and function of context principle in Frege's thought. Resnik (1980) seems to be still the best place to start this line of research, and Hallett (2010, 2012) must be helpful for a quick update. There is also another trend in Frege scholarship, i.e., studies on Frege's views on thinking, which may give further support to all this (for example, Garavaso and Vassallo 2015). For, Frege's views on concept formation in mathematics can be better understood in broader perspective by studying his views on thinking and thought without unnecessary worries about psychologism.

My study of the Hilbert School is bound to incomplete in many ways. If my study of Zermelo or Bernays has some valuable elements, the similar approaches to Husserl or Weyl could be also worthwhile to pursue. In fact, we can witness exactly that in some recent studies. There is a huge literature dealing with Husserl's early work in logic and mathematics, which must be properly understood only by understanding the Hilbertian heritage in Husserl (Hill and Da Silva 2013). D. Christopoulou's recent study of Weyl's views on implicit definitions is particularly encouraging, for this seems to demonstrate that my basic points are on the right track (Christopoulou 2014). Corry (2004) is essential for widening the scope of studying Hilbert and the Hilbertian heritage even further. The publication of Hilbert's lecture notes is good news. Of course, we have reason to expect the early completion of Bernays project and Carnap project.

My discussion of implicit definition in the context of logical positivism shows some possible benefits of extending the discussion of philosophy of mathematics to that of philosophy of science. In this regard, Stöltzner's recent researches seem to provide us with a nice point of departure. In particular, the question raised by Stöltzner (2002), i.e., 'How metaphysical is "deepening the foundations"?' can be extended meaningfully to all the members of logical positivism as well as Hahn and Frank. Stöltzner (2015) is a clear example that could be instrumental to understanding the history of metalogic from the interaction between the Hilbert School and logical positivism. Recent renaissance in Carnap seems quite fortunate from this point of view (e.g., Wagner 2009, 2012). Especially, the recent scholarship on Carnap's early semantics seems quite impressive (Schiemer 2012, 2013; Loeb 2014).

The chapters in Part 3 can also be the starter for further fruitful studies in history of logic. As is clear from my review of Patterson (2012) in Chap. 8, the Polish tradition in the twentieth century logic is extremely important. Fortunately, we can witness the sustained efforts to uncover the tradition (Patterson 2007; Mulligan et al. 2014). It is certainly mandatory for any historian of logic, mathematics, and science to scrutinize the interactions between logicians in Vienna, Prague, Berlin, Göttingen, and Warsaw. However, it seems to me the uniqueness of the Polish logic may be better captured by the Aristotelian heritage embedded in Tarski's logic teachers. Lesniewski's views on the axiomatic method, for example, could be of utmost interest. Tarski's struggle with his teachers' long lasting influence can be detected in many areas of thought, especially metaphysics.

Szatkowski (2012) is an impressive collection of essays on the ontological proof of the existence of God. This is evidently inspired by Gödel, showing the utmost sophistication of modern logic. What is surprising is that it is hard to find answers to my question as to what motivated Gödel to draft such a proof. Probably, much more relevant to my purpose is Gierer (1997), which deals with an interesting dialogue between Gödel and Carnap on God and religion. It could be a nice working historical conjecture that the dialogue was the immediate cause of Gödel's ontological proof. Be that as it may, I am quite convinced that the Gödel's ontological proof presents us ample data for how he understood the axiomatic method. However stupid it might appear to the experts, I think, we should start by raising question such as "Is the term 'God' supposed to be defined, as Gödel did?", "Was Gödel assuming that the term 'God' on a par with the theoretical terms in science?", or "By alluding to the possibility of proving the existence of God, was Gödel launching a *reductio ad absurdum* argument against Carnap?" The idea of treating the historically extant ontological proofs as the source for fathoming how axiomatic method was understood by particular philosophers or scientists seems quite promising. In this vein, for example, we may ask what a Newtonian ontological proof might look like.

As was already confessed, I discussed Maddy's and Shapiro's philosophies of mathematics, primarily because they seem to be closer to the ideal philosophy of mathematics for the future, i.e., Aristotelian philosophy of mathematics. In Maddy's more recent work, I cannot find any attempt to develop the Aristotelian elements even further (Maddy 2007, 2011; see also Roland 2007, 2009). Nor can I find the explosion of scholarly interest in Shapiro/Keränen controversy. On the other hand, It is indeed encouraging to find that Aristotelian philosophy of mathematics is becoming more and more influential. James Franklin deserves a special credit for this unexpected development (Franklin 2009, 2011, 2012, 2014, 2015, forthcoming). As Jacquette concurs, Franklin's complaint on ignoring the Aristotelian option is fully justified:

> To pose the problem in those terms [Platonism versus nominalism] neglects the Aristotelian option: that mathematics may be about some real properties of the (physical and/or non-physical) world, such as its quantitative properties, or its symmetry, continuity, structure or pattern. Such a view is neither Platonist (since those properties are not in an "abstract" world but realized or at least realizable in the actual world, and in simple cases are perceivable) nor nominalist (since those properties are real aspects of things). Despite some Aristotelian currents in certain authors, there is no recognized complete Aristotelian option or school in the philosophy of mathematics. (Franklin 2011, 3)

At first blush, establishment of the Aristotelian option appears to be an arduous task. For, as Jacquette points out,

> To make progress toward a viable Neoaristotelian alternative that can credibly oppose classical default, dyed-in-the wool and deeply entrenched Platonism in the philosophy of symbolic logic and mathematics, it is necessary to : (1) explain the Aristotelian concept of the inherence of secondary substances in primary substances, and hence of mathematical properties embodied and embedded in primary substances (physical entities); (2) identify a special category for ostensible mathematical formalism that are independent of the inherence of any mathematical properties in any physical entities or Aristotelian primary substances. (Jacquette 2014, 168)

There is no doubt that setting up such desiderata for any future Aristotelian philosophy of mathematics is to be heartily welcomed. Nevertheless, I tend to feel that Franklin and Jacquette might be forgetting one thing. They seem to be talking about Aristotelian philosophy of mathematics as a non-existent possible. In fact, there have been Aristotelian philosophies of mathematics throughout the history of mathematics. My discussion of Biancani's views on *scientiae mediae* is a clear historical example. This was inspired by Mancosu (1996), which still remains to be the best introduction to the history of philosophy of mathematics. In other words, what is needed is the any number of careful historical case studies on Aristotelian mathematicians and scientists throughout the history.

In this regard, I may draw the readers' attention to Cocchiarella's study of medieval supposition theory (Cocchiarella 2007). I believe that his study of the theories of universals (Cocchiarella 1986) is the crowning achievement in the history of ontology. His study of early analytic philosophy (Cocchiarella 1987) is the model of doing philosophy both logically and historically. It is significant that he included at least two chapters on medieval logic in his magnum opus (Cocchiarella 2007). It is unfortunate that, to the best of my knowledge, no leading scholar in medieval logic has discussed Cocchiarella's contribution to medieval theory of suppositio extensively. I sincerely hope to read in the near future articles dealing with Cocchiarella's reinterpretation of medieval supposition theory or his refutation of Geach's classical arguments (see also Geach 1962, 1980; Klima 2015; Orenstein 2015). In my case, it has not been easy to understand Ockham's position, not to mention the elements of conceptualism in Cocchiarella's formal ontology. It seems, however, highly promising to reconstruct Cocchiarella's Aristotelian conceptualist philosophy of mathematics from his scattered remarks on arithmetic. If I am right, we may find an instance of an Aristotelian philosophy of mathematics meeting Jacquette's desiderata as well as my interest in historical cases in Cocchiarella's study of medieval logic.

Finally, I would like to emphasize the obvious connection of the present work with my recent book on abduction (Park 2016). Charles S. Peirce might be the one who anticipated all my interests expressed in this book. He was one of the pioneers of modern logic. But he was at the same time the first rate Aristotle scholar. It is also well known that Scotus and Ockham were two of the greatest metaphysicians from his point of view. In a chapter on Peirce's and Magnani's views on geometrical diagrams, I reconstructed an imaginary dialogue between Charles S. Peirce and his father, Benjamin Peirce, on the nature of mathematics (Park 2016). In some sense, not only that chapter but also the entire volume could be read as a continuation of my study presented in this monograph. It is rather tempting to rehearse the dialogue between Peirces in order to make the connection clearer. Let it suffice to show you an abstract of a paper I am currently working on.

[Abstract] On Abducing the Axioms of Mathematics

In view of the importance of mathematics to Charles S. Peirce's life and work, it is a scandal of philosophy that it has failed to grant a fair hearing to the role and function of abduction in mathematics. Of course, there are some notable exceptions such as Lorenzo Magnani's discussion of Gödelian abduction (Magnani 2009). or Dov Gabbay and John Woods' discussion of regressive abduction (Gabbay and Woods 2005). However, it seems rather obvious that mathematical abduction would be a red herring to most practicing mathematicians. In order to rectify all this, I shall attempt two ambitious projects. First, I shall examine critically some recent studies on the origin of the axioms of mathematics. Second, I will look into Peirce's writings in order to fathom how he would have responded to the issues raised by the examination.

Here is my strategy to approach these two projects. In Sect. 1 of my forthcoming paper, I will discuss Thomas Forster's important paper "The Axiom of Choice and the Inference to the Best Explanation" (Forster 2006). I will find fault with his identification of abduction with IBE only to highlight the significance of abduction in building axiomatic systems. In Sect. 2, I shall diagnose how it was possible for Magnani to characterize Gödel's infamous claims on mathematical intuition in terms of abduction. In Sect. 3, I will compare Magnani's interpretation with that of Charles Parsons, who has shown continued interest in the Gödelian intuition of the mathematicals (Parsons 2014). In Sect. 4, by following closely Gabbay and Woods' guide, I will discuss the problem of mathematical abduction in broader perspective, i.e., the explanatory versus non-explanatory abduction. In Sect. 4, there will be brief digression to the problem of mathematical explanation. For, though it is rather a new comer to philosophy of mathematics, largely thanks to Paolo Mancosu, it may have far-reaching implications to the entire history of scientific method (Mancosu 2001). At least, this digression could be useful for sharpening the contrast between mathematical abduction and mathematical IBE. In Sect. 5, I will present a summary of what I elsewhere reconstructed as the imaginary dialogue between Charles and Benjamin Peirce on mathematical reasoning. Finally, in Sect. 6, I will turn to Peircean corpus to learn how he would have replied to some of the ideas discussed in earlier sections. Hopefully, by then, we will be able to return to our point of departure, i.e., the axiom of choice and Peirce's stance, if any, to it, which must be an excellent test for all wild speculations in my paper.

References

Bernays, P. (1942). Review of Max Steck, 'Ein Unbekannter Brief von Gottlob Frege uber Hilberts serste Vorlesung uber die Grundlgen der Geometrie'. *Journal of Symbolic Logic, 7*(2), 92–93.

Christopoulou, D. (2014). Weyl on Fregean implicit definitions: Between phenomenology and symbolic construction. *Journal of General Philosophy of Science, 45,* 35–47.

Cocchiarella, N. B. (1986). *Logical investigations of predication theory and the problem of universals.* Napoli: Bibliopolis.

Cocchiarella, N. B. (1987). *Logical studies in early analytic philosophy.* Columbus, OH: Ohio State University Press.

Cocchiarella, N. B. (2007). *Formal ontology and conceptual realism.* Dordrecht: Springer.

Cook, R. T. (Ed.). (2007). *The Arché papers on the mathematics of abstraction.* New York: Springer.

Corry, L. (2004). *David Hilbert and the axiomatization of physics (1898–1918.* Dordrecht: Kluwer.

Ebert, P. A., & Rossberg, M. (2016). *Abstractionism: Essays in philosophy of mathematics.* Oxford: Oxford University Press.

Forster, T. (2006). The axiom of choice and the inference to the best explanation. *Logique et Analyse, 49,* 191–197.

Franklin, J. (2009). Aristotelian realism. In A. D. Irvine (Ed.), *Philosophy of mathematics* (pp. 103–155). Amsterdam: Elsevier.

Franklin, J. (2011). Aristotelianism in the philosophy of mathematics. *Studia Neoaristotelica, 8,* 3–15.

Franklin, J. (2012). Science by conceptual analysis: The genius of the late scholastics. *Studia Neoaristotelica, 9,* 3–24.

Franklin, J. (2014). *An Aristotelian realist philosophy of mathematics: Mathematics as the science of quantity and structure.* London: Palgrave Macmillan.

Franklin, J. (2015). Uninstantiated properties and Semi-Platonist Aristotelianism. *The Review of Metaphysics, 69,* 25–45.

Franklin, J. (forthcoming). *Varieties of Aristotelian realism in the philosophy of mathematics.* Not yet published.

Gabbay, D. M., & Woods, J. (2005). *The reach of abduction: Insight and trial.* Amsterdam: Elsevier.

Garavaso, P., & Vassallo, N. (2015). *Frege on thinking and its epistemic significance.* Lanham, Maryland: Lexington Books.

Geach, P. (1962, 1980). *Reference and generality* (3rd Ed.). Ithaca: Cornell University Press.

Gierer, A. (1997). Gödel meets Carnap: A prototypical discourse on science and religion. *Zygon, 32*(2), 207–217.

Hale, B., & Wright, C. (2001). *The reason's proper study: Essays towards a Neo-Fregean philosophy of mathematics.* Oxford: Clarendon Press.

© Springer International Publishing AG, part of Springer Nature 2018

W. Park, *Philosophy's Loss of Logic to Mathematics*, Studies in Applied Philosophy, Epistemology and Rational Ethics 43, https://doi.org/10.1007/978-3-319-95147-8

Hallett, M. (2010). Frege and Hilbert. In M. Potter & T. Ricketts (Eds.), *The Cambridge companion to Frege* (pp. 413–464). Cambridge: Cambridge University Press.

Hallett, M. (2012). More on Frege and Hilbert. In M. Frappier, et al. (Eds.), *Analysis and interpretation in the exact sciences* (pp. 135–162). Dordrecht: Springer.

Hill, C. O., & Da Silva, J. J. (2013). *The road not taken: On Husserl's philosophy of logic and mathematics*. Milton Keynes, UK: College Publications.

Jacquette, D. (2014). Toward a Neoaristotelian inherence philosophy of mathematical entities. *Studia Neoaristotelica, 11,* 159–204.

Klima, G. (2015). Geach's three most inspiring errors concerning medieval logic. *Philosophical Investigations, 38*(1–2), 34–51.

Loeb, I. (2014). Uniting model theory and the universalist tradition of logic: Carnap's early axiomatics. *Synthese, 191,* 2815–2833.

Maddy, P. (2007). *Second philosophy. A naturalistic method.* New York: Oxford University Press.

Maddy, P. (2011). *Defending the axioms: On the philosophical foundations of set theory.* New York: Oxford University Press.

Magnani, L. (2009). *Abductive cognition: The epistemological and eco-cognitive dimensions of hypothetical reasoning.* Berlin: Springer.

Mancosu, P. (1996). *Philosophy of mathematics and mathematical practice in the Seventeenth Century.* Oxford: Oxford University Press.

Mancosu, P. (2001). Mathematical explanation: Problems and prospects. *Topoi, 20,* 97–117.

Orenstein, A. (2015). Geaach, Aristotle and predicate logics. *Philosophical Investigations, 38*(1–2), 96–114.

Mulligan, K., et al. (Eds.). (2014). *The history and philosophy of polish logic essays in honour of Jan Woleński.* UK: Palgrave Macmillan.

Park, W. (2016). *Abduction in context: The conjectural dynamics of scientific reasoning.* Dordrecht: Springer.

Park, W. (forthcoming). *Abducing axioms of mathematics.*

Parsons, C. (2014). *Philosophy of mathematics in the twentieth century.* Cambridge, MA: Harvard University Press.

Patterson, D. (Ed.). (2007). *New essays on Tarski and philosophy.* Oxford: Oxford University Press.

Patterson, D. (2012). *Alfred Tarski: Philosophy of language and logic.* Palgrave Macmillan.

Resnik, M. (1980). *Frege and the Philosophy of Mathematics.* Ithaca and London: Cirnell University Press.

Robinson, R. (1963). *Definition.* Oxford: Oxford University Press.

Roland, J. W. (2007). Maddy and mathematics: Naturalism or not. *British Journal for the Philosophy of Science, 58,* 423–450.

Roland, J. W. (2009). On naturalizing the epistemology of mathematics. *Pacific Philosophical Quarterly, 90,* 63–97.

Schiemer, G. (2012). Carnap's Untersuchungen: Logicism, formal axiomatics, and metatheory. In R. Creath (Ed.), *Rudolf Carnap and the legacy of logical empiricism* (pp. 13–36). Dordrecht: Springer.

Schiemer, G. (2013). Carnap's early semantics. *Erkenntnis, 78,* 487–522.

Szatkowski, M. (Ed.). (2012). *Ontological proofs today.* Frankfurt: Ontos Verlag.

Stöltzner, M. (2002). How metaphysical is "deepening the foundations? Hanhn and Frank on Hilbert's axiomatic method. In M. Heidelberger & F. Stadler (Eds.), *History and philosophy of science* (pp. 245–262). Dordrecht: Kluwer.

Stöltzner, M. (2015). Hilbert's axiomatic method and Carnap's general axiomatics. *Studies in History and Philosophy of Science, 53,* 12–22.

Wagner, P. (Ed.). (2009). *Carnap's logical syntax of language.* UK: Palgrave Macmillan.

Wagner, P. (Ed.). (2012). *Carnap's ideal of explication and natralism.* UK: Palgrave Macmillan.